Sustainable Production, Life Cycle Engineering and Management

Series Editors

Christoph Herrmann, Braunschweig, Germany

Sami Kara, Sydney, Australia

SPLCEM publishes authored conference proceedings, contributed volumes and authored monographs that present cutting-edge research information as well as new perspectives on classical fields, while maintaining Springer's high standards of excellence, the content is peer reviewed. This series focuses on the issues and latest developments towards sustainability in production based on life cycle thinking. Modern production enables a high standard of living worldwide through products and services. Global responsibility requires a comprehensive integration of sustainable development fostered by new paradigms, innovative technologies, methods and tools as well as business models. Minimizing material and energy usage, adapting material and energy flows to better fit natural process capacities, and changing consumption behaviour are important aspects of future production. A life cycle perspective and an integrated economic, ecological and social evaluation are essential requirements in management and engineering.

Indexed in Scopus

To submit a proposal or request further information, please use the PDF Proposal Form or contact directly: Petra Jantzen, Applied Sciences Editorial, email:petra.jantzen@springer.com

More information about this series at http://www.springer.com/series/10615

Stefan Albrecht · Matthias Fischer ·
Philip Leistner · Liselotte Schebek
Editors

Progress in Life Cycle Assessment 2019

Springer

Editors
Stefan Albrecht
Life Cycle Engineering GaBi
Fraunhofer Institute for Building Physics
Stuttgart, Baden-Württemberg, Germany

Matthias Fischer
Life Cycle Engineering GaBi
Fraunhofer Institute for Building Physics
Stuttgart, Baden-Württemberg, Germany

Philip Leistner
Fraunhofer Institute for Building Physics
Stuttgart, Baden-Württemberg, Germany

Liselotte Schebek
Material Flow Management and Resource Economy
Technical University of Darmstadt
Darmstadt, Hessen, Germany

ISSN 2194-0541 ISSN 2194-055X (electronic)
Sustainable Production, Life Cycle Engineering and Management
ISBN 978-3-030-50518-9 ISBN 978-3-030-50519-6 (eBook)
https://doi.org/10.1007/978-3-030-50519-6

This Springer imprint is published by the registered company Springer Nature Switzerland AG
The registered company address is: Gewerbestrasse 11, 6330 Cham, Switzerland

Contents

Chapter 1
Cooperation of Young Researchers from Science and Industry — Life Cycle Assessment in Theory and Practice

Stefan Albrecht and Matthias Fischer

Abstract This preface introduces the contributions from young scientists and researchers participating in the 15th Ökobilanzwerkstatt in Stuttgart. It provides an overview of the growing relevance of establishing environmental aspects as a guideline in science, industry and politics. This results in an increased need to anchor life cycle assessments in the training of young scientists. Afterwards, the highlights of the Ökobilanzwerkstatt 2019 are presented and the topics of the publication in this book are introduced.

Keywords Life cycle assessment · Life cycle engineering · Ökobilanzwerkstatt

1.1 Introduction

The need to actively address current environmental issues is more present than ever. The discussion about the limitation of climate change, the consequences of climate change that are already being felt in many places, such as the locally distributed increase in extreme weather phenomena such as heavy rainfall and droughts, as well as the increasing competition for raw materials such as cobalt, sand or water, illustrate this impressively.

Looking ahead to the future, politicians and companies are increasingly gearing their strategies to environmental guidelines. The 17 global goals for sustainable development of Agenda 2030, the "Sustainable Development Goals (SDG)" of the United Nations, are aimed at everyone: governments worldwide, but also civil society, the private sector and the scientific community (The Sustainable Development Goals knowledge platform of the United Nations UN 2020). The European Commission presented the Green Deal on 11 December 2019. It comprises a package of legislative and policy initiatives to implement a new growth strategy and achieve climate

---

S. Albrecht (✉) · M. Fischer
Life Cycle Engineering GaBi, Fraunhofer Institute for Building Physics IBP, 70563 Stuttgart, Germany
e-mail: stefan.albrecht@ibp.fraunhofer.de

S. Albrecht et al. (eds.), *Progress in Life Cycle Assessment 2019*,
Sustainable Production, Life Cycle Engineering and Management,
https://doi.org/10.1007/978-3-030-50519-6_1

neutrality by 2050. European policy aims to make Europe the first climate-neutral continent by that date.

The industrialized societies, which include the German-speaking countries, have a special role to play here in implementing environmental and climate protection measures. On the one hand, they make a significant contribution to global emissions, and on the other hand, as exporters of environmentally and climate-friendly technologies, they have the role of supporting and supplying emerging countries with such technologies.

With the method of life cycle assessment, an instrument was already created at the beginning of the 1990s which makes it possible to support research and development of environmentally friendly products in a targeted manner. The life cycle assessment has established itself worldwide as a standard instrument for the analysis, evaluation and assessment of products, processes and services. „In its Communication on Integrated Product Policy (COM (2003)302), the European Commission concluded that Life Cycle Assessments provide the best framework for assessing the potential environmental impacts of products currently available“ (European Platform on Life Cycle Assessment (LCA) 2020).

1.2 The Need for Young Scientists in the Field of Life Cycle Assessment

The more science, industry and society orientate themselves towards the guidelines of environmentally sound and sustainable development, the greater the need for well and comprehensively trained natural scientists, engineers, economists, etc., who further develop established methods and take into account and implement the findings in decision-making processes. In 2005 the Ökobilanzwerkstatt took place for the first time as an event for young scientists who are doing research in the field of life cycle analysis or who use life cycle analysis for applied questions. It is intended to provide a forum for young scientists to present scientific developments and new methods as well as for scientific discussion of their work and for personal exchange of experience. Baitz et al. discussed the necessity of LCA in theory and practice as to "share the implications of LCA in daily businesses and practice and aim to nurture and strengthen the interfaces between scientific findings and application (Baitz et al. 2013)".

1.3 The Ökobilanzwerkstatt in Stuttgart—Life Cycle Assessment in Theory and Practice

According to the motto "Life Cycle Assessment in Theory and Practice", the 15th Ökobilanzwerkstatt was successfully organized in Stuttgart in 2019 by the Fraunhofer Institute for Building Physics IBP, Life Cycle Engineering GaBi.

The close networking of research and industry, the guiding principle of the Ökobilanzwerkstatt 2019, is indispensable in the field of life cycle assessment and has been the daily practice of the department of Life Cycle Engineering GaBi for more than 30 years.

About 60 participants from Germany and Austria discussed the successful application of life cycle assessment in their daily work, the high data quality as a prerequisite for transfer to different fields of sustainability, but also new scientific approaches. The presentations addressed a wide range of current topics, from mobility and aviation, through circular economy and life cycle assessments in industry, to sustainable construction. Research on resource efficiency and resource availability, biotechnology, recycling processes and global waste streams as well as future developments and challenges were also discussed. Last but not least, presentations on impact assessment in the fields of toxicity, biodiversity, water footprint and on social and societal challenges illustrated the broad field of application of life cycle analyses.

Prof. Dr. Liselotte Schebek from the Technical University of Darmstadt, who has been the patron of the Ökobilanzwerkstatt since it was first held in 2005, gave an insight into its history and underlined its importance for young scientists. The LCA workshop was rounded off by impulse lectures from industry. Dr. Martin Baitz from Sphera Solutions GmbH (formerly Thinkstep AG) gave a keynote speech entitled "LCA as a profession! Why? What? Where?". He gave exciting insights into the working environment of LCA users in industry. He impressively underlined the importance of life cycle assessments in many companies and the high relevance of a sound life cycle assessment training and the good career prospects of life cycle assessment experts. Dr. Anna Braune from the German Sustainable Building Council (Deutsche Gesellschaft für nachhaltiges Bauen DGNB) focused in her presentation on the relevance of life cycle assessments for sustainable building. She explained how life cycle assessments support and inspire sustainable building and provide evidence for environmentally friendly building. In addition, Prof. Dr. Jan Paul Lindner (Bochum University of Applied Sciences and Fraunhofer IBP, Life Cycle Engineering GaBi) and Prof. Dr. Tobias Viere (Pforzheim University of Applied Sciences) enriched the LCA workshop with valuable contributions, discussions and suggestions regarding content. Last but not least, there were many good opportunities for personal exchange and networking during the event.

1.4 Progress in Life Cycle Assessment 2020

The articles published in this book were written by participants of the Ökobilanzwerkstatt 2019 in Stuttgart. The articles show the current state of work and research of the respective authors. Some of the papers represent early stages, some more advanced stages of their research. This book covers the latest developments in life cycle assessment LCA both in terms of methodology and its application in various research areas, including mobility, technology and production. With numerous research articles from leading German research institutes, the book provides an informative source for professionals working in the field of sustainability assessment, researchers interested in the current state of LCA research, and advanced university students in various scientific and technical fields.

We are pleased to present this collection of high-class scientific contributions as a compilation of selected results of the Ökobilanzwerkstatt 2019 in this book.

Stefan Albrecht and Matthias Fischer.

References

Baitz M, Albrecht S, Brauner E et al (2013) LCA's theory and practice: like ebony and ivory living in perfect harmony? Int J Life Cycle Assess 18:5–13. https://doi.org/10.1007/s11367-012-0476-x

European Platform on Life Cycle Assessment (LCA). https://ec.europa.eu/environment/ipp/lca.htm. Last accessed 2020/04/30

The Sustainable Development Goals knowledge platform of the United Nations UN. https://sustainabledevelopment.un.org. Last accessed 2020/04/30

Chapter 2
Life Cycle Assessment of a Hydrogen and Fuel Cell RoPax Ferry Prototype

Juan Camilo Gomez Trillos, Dennis Wilken, Urte Brand, and Thomas Vogt

Abstract Estimates for the greenhouse gas emissions caused by maritime transportation account for approx. 870 million tonnes of CO_2 tonnes in 2018, increasing the awareness of the public in general and requiring the development of alternative propulsion systems and fuels to reduce them. In this context, the project HySeas III is developing a hydrogen and fuel cell powered roll-on/roll off and passenger ferry intended for the crossing between Kirkwall and Shapinsay in the Orkney Islands in Scotland, a region which currently has an excess of wind and tidal power. In order to explore the environmental aspects of this alternative, a life cycle assessment from cradle to end-of-use using the ReCiPe 2016 method was conducted, contrasting the proposed prototype developed within the project against a conventional diesel ferry and a diesel hybrid ferry. The results show that the use of hydrogen derived from wind energy and fuel cells for ship propulsion allow the reduction of greenhouse gas emissions of up to 89% compared with a conventional diesel ferry. Additional benefits are lower stratospheric ozone depletion, ionizing radiation, ozone formation, particulate matter formation, terrestrial acidification and use of fossil resources. In turn, there is an increase in other impact categories when compared with diesel electric and diesel battery electric propulsion. Additionally, the analysis of endpoint categories shows less impact in terms of damage to human health, to the ecosystems and to resource availability for the hydrogen alternative compared to conventional power trains.

Keywords Hydrogen · Fuel cells · LCA · Shipping emissions reduction

---

The original version of this chapter was revised: The legend of the figure 2.2 was corrected. The correction to this chapter is available at https://doi.org/10.1007/978-3-030-50519-6_17

J. C. G. Trillos (✉) · U. Brand · T. Vogt
DLR Institute of Networked Energy Systems, Carl-von-Ossietzky-Straße 15, 26129 Oldenburg, Germany
e-mail: juan.gomeztrillos@dlr.de

D. Wilken
Stedinger Straße 43a, 26135 Oldenburg, Germany
e-mail: wilken.de@gmail.com

S. Albrecht et al. (eds.), *Progress in Life Cycle Assessment 2019*,
Sustainable Production, Life Cycle Engineering and Management,
https://doi.org/10.1007/978-3-030-50519-6_2

2.1 Introduction

According to the "Third IMO GHG Study 2014", international shipping emitted 796 million tonnes of CO_2, or considering other emissions, 816 million tonnes of CO_2eq in 2012 and represented approximately 2.1% of the global CO_2eq emissions in the same year (Smith et al. 2015). Estimates for 2018 amount to 870 million tonnes of CO_2 (DNV GL–Maritime 2019). Moreover, an increase of between 50% and 250% of CO_2 is expected on a business-as-usual scenario by 2050 (Smith et al. 2015). This has led to efforts in reducing the sector's greenhouse gas (GHG) emissions. Consequently, the Marine Environment Protection Committee (MEPC) of the IMO issued the Resolution MEPC.304(72) in 2018, adopting an initial strategy for reducing GHG emissions, aiming to reduce them by 50% by 2050 using share of emissions in 2008 as reference level. This resolution describes short, medium and long term measures for this purpose. Within this resolution, short term measures are related with energy efficiency and regulations; however, the implementation of zero-carbon or fossil-free fuels together with emission reduction mechanisms are the only measures envisioned in the long term to reduce GHG emissions (International Maritime Organization 2018).

So far, the strategy of the European Union in this regard consists of monitoring, reporting and verification in the first place, establishing greenhouse gas reduction targets in the second place and further measures including market-based measures as a last step (European Commission 2019). The first step already commenced with the obligation from January 1st 2018 for large ships over 5000 gross tonnage to monitor and report their CO_2 emissions when loading or unloading cargo or passengers at ports in the European Economic Area (EEA) (European Commission 2019). However, another strategy paired with the long term ambition of reducing or decarbonising the shipping sector is the development of low carbon fuels and alternative power and propulsion systems for ships.

2.1.1 Ferries and Project HySeas III

Ferries are ships conveying passengers and goods, especially over a relatively short distance and as a regular service.[1] Roll on/roll-off/Passenger (RoPax) ferries have the special feature of being designed to carry wheeled payload, particularly vehicles, and passengers. According to numbers from the database SeaWeb IHS Markit, 42% of the RoPax ferry ships listed globally are in European registers, meaning that approximately 1400 operate in Europe (IHS Markit 2019). Around 40% of the fleet is more than 30 years old (IHS Markit 2019). The average lifetime of ferries is around 35 years, meaning that many of these ships will require replacement or retrofitting in the near future to continue with the transportation services.

[1] https://www.lexico.com/en/definition/ferry.

In this context, the current research project HySeas III (2018–2021) has the motivation to realise the first sea-going RoPax ferry powered by hydrogen and fuel cells (HySeas III Project 2019). The ship will be 40 m long, 10 m wide, and will have a capacity of 120 passengers and 20 passenger vehicles or 2 lorries. It is intended that the ship will operate at the crossing between Kirkwall and Shapinsay in the Orkney Islands, Scotland. The sister projects BIG HIT and SURF 'N' TURF are already developing a hydrogen supply chain at this location, including production via electrolysers, storage, transportation in high pressure tanks and applications linked to the use of hydrogen (BIG HIT Project 2019; Surf 'N' Turf Project 2019). The production of hydrogen will be mainly based on wind power and tidal power available at the location, and could supply the ship developed by project HySeas III in the near future. Moreover, the use of hydrogen has been envisioned as an option for the routes Barra–Eriskay and Stornoway–Ulapool also in Scotland (Point and Sandwick Trust 2019).

In addition to the technical development of a hydrogen-powered fuel cell RoPax ferry prototype, HySeas III aims to assess the economic, environmental and social impacts of the particular application under development. The main driver is the reduction of the environmental impact of ships. Thus, the HySeas III consortium decided to conduct an environmental assessment of the prototype considering different aspects surrounding the project. Life cycle assessment (LCA) is a methodology that allows consideration of environmental aspects and potential environmental impacts throughout the life cycle of a product (DIN Deutsches Institut für Normung e.V., Umweltmanagement – Ökobilanz – Grundsätze und Rahmenbedingungen (ISO 14040). Therefore, this methodology was selected to perform the environmental analysis of the ship to be built by HySeas III.

2.1.2 Previous Approaches of Life Cycle Assessment of Alternative Fuels for Ships

Several authors have previously approached the topic of life cycle assessment of alternative fuels for ships. Gilbert et al. analysed different fuels and production paths considering the impacts on three greenhouse gases (CO_2, CH_4 and N_2O) and three local pollutants (SO_x, NO_x and PM) (Gilbert et al. 2018). Liquid hydrogen was found as the best alternative in terms of reducing GHG emissions and local pollutants, but only if it is produced by the use of renewable energy and particularly of wind power. This study also served to demonstrate the applicability in practice and upscaling up to industrial level. Nevertheless, this study only considered the life cycle of energy carriers, leaving aside the life cycle of the equipment.

A bachelor thesis performed by Jokela et al. conducted a life cycle assessment of a hydrogen-electric ferry, finding a reduction of 79% of the global warming potential (GWP) emissions of the assessed alternative compared to a conventional diesel ferry (Jokela et al. 2019). This analysis considered the ferry connection Hjelmeland–Skipavik–Nesvik in Norway. In that case, 50% of the electricity used for ship propulsion must come from hydrogen, while the remaining 50% shall be electricity from on-board batteries, which are charged at the docking sites. The functional unit in this study was the 3 km crossing between Hjelmeland and Nesvik, consisting of 13 min crossing time and 7 min at the quay. Moreover, the authors assumed a fuel cell power system of 746 kW and batteries with a capacity of 746 kW. After using the CML-IA impact assessment method, the authors calculated an impact of 5.2 kg CO_2eq/crossing for the hydrogen-electric ferry and 24.8 kg CO_2eq/crossing for the diesel ferry (Jokela et al. 2019). In comparison to the ferry analysed by Jokela et al. the ferry developed in HySeas III would have a different operation profile and specifications, and hydrogen would be produced using different electricity sources.

Other LCAs for electric vehicles and fuel cell vehicles have shown the relevance of taking the equipment or even the infrastructure into account (Nordelöf et al. 2014; Marmiroli et al. 2019). Although the impact of these components in terms of GWP might be low, their impact in other impact categories can be relevant, and therefore additional impact categories were highlighted in this study.

2.2 Methodology

The methodology of life cycle assessment (LCA) in compliance with the ISO 14040 and ISO 14044 standards was employed to estimate the environmental impact of different ship alternatives (DIN Deutsches Institut für Normung e.V., Umweltmanagement – Ökobilanz – Grundsätze und Rahmenbedingungen (ISO 14040; DIN Deutsches Institut für Normung e.V., Umweltmanagement – Ökobilanz – Anforderungen und Anleitungen (ISO 14044). The next section describes the application of the methodology according to the four basic steps required by these standards for conducting an LCA study: goal and scope definition, inventory analysis, impact assessment and interpretation of the results.

2.2.1 Goal

The main goal of this study was to conduct an environmental assessment of the proposed hydrogen powered fuel cell RoPax ferry developed within the project HySeas III, which is to be implemented on the route Kirkwall–Shapinsay. This alternative was compared with other conventional propulsion systems, including diesel engines and diesel hybrid systems in conjunction with on-board batteries operating on the same route. The aim of this analysis was to establish the benefits and drawbacks of using hydrogen and fuel cells for this transportation service.

The intended audience for this study are decision-makers related to the maritime sector, industry, politicians, scientists, companies and the public in general. The background data used in the study is based on ecoinvent 3.5 (Wernet et al. 2016) and the software utilized for the calculations was SimaPro 9.0. Primary data in terms of modelled energy consumption and component sizing was collected from the project partners, particularly those involved with the construction and trials of the power system (Ferguson, Ballard, Kongsberg) and with the operation of current ferries and of the future prototype (Orkney Council–Marine Services). The information gaps not fulfilled by project partners or protected as part of their business secrets were covered using literature sources as described in the next sections.

2.2.2 Scope

The main service provided by the operation of RoPax ferries is the transportation of passengers and vehicles. In order to perform this task, different elements are necessary. Within this analysis, an on-board power source to propel the ship and at the same time cover the internal energy demand of the different systems, a storage system of fuel carriers on-board, a dispensing unit to load the fuel on the storage system of the vessel and the upstream supply chain of the energy carrier used during this operation were considered. These were taken into account from cradle to end-of-use, meaning that the final disposal of the ship and its components were not considered for this approach. Other elements such as the quayside facilities employed as link between the ship and the mainland and the personnel involved in the operation of the ship were not taken into account in the analysis and therefore are considered out of the scope.

Considered Ship Systems and Functions

According to Papanikolaou, a ship's design involves considerations for a variety of subsystems serving a series of functions, divided into payload functions and ship inherent functions (Papanikolaou 2019). The payload functions are related to the provision of cargo (passengers) spaces, cargo handling and cargo treatment, whereas the ship inherent functions involve the carriage or transport of cargo or in other words the hull, superstructures and a propulsion/power unit together with fuel to enable the transport from A to B (Papanikolaou 2019). The main reason for choosing these subsystems is that these are the ones that would differ between different propulsion alternatives, as would be the case of a hydrogen powered fuel cell ferry.

Table 2.1 shows a summary of the inherent and payload functions. This work focusses on inherent functions, including structure, machinery and tanks, which themselves are comprised of the listed subsystems. and among them underlined subsystems. Only those underlined functions were considered within the scope of this work. Those functions are performed by different components in the ship, as will be described in the inventory analysis section. The structure was included to estimate the impact of the hull material in comparison to other components. Moreover, the machinery and tanks were included because they constitute the different propulsion systems and vary between the considered alternatives. The other functions were

Table 2.1 Ship functions

Inherent function		Payload function	
Structure	Hull, poop deck, forecastle superstructures	Cargo units	Containers, trailers, cassettes, pallets, bulk/break bulk
Crew facilities	Crew spaces, Service spaces, stairs and corridors	Cargo spaces	Holds, deck cargo spaces, cell guides, tanks
Machinery	Engine and pump rooms, engine casing, funnel, steering and thrusters	Cargo handling	Hatches and ramps, cranes, cargo pumps, lashing
Tanks	Fuel and lub oil, water and sewage, ballast and voids	Cargo treatment	Ventilation, heating and cooling, pressurising
Comfort systems	Air conditioning, water and sewage		
Outdoor decks	Mooring, lifeboats, etc.		

Adapted from (Papanikolaou 2019)

not considered here, as they would not diverge considerably among the different alternatives, and due to lack of information at this design stage.

Functional Unit

The functional unit (FU) used for this study was 1 km of crossing distance of the selected ship during the lifetime considered for the purposes of this study of 30 years. The lifetime of the ships was selected according to the lifetime described for previous RoPax ferries by databases such as SeaWeb from the company IHS Markit, given that most of the ships reach at least this lifetime. Furthermore, the selection of the functional unit in terms of distance was done because the RoPax ferries have the two main functionalities of transporting passengers and transporting vehicles, which would require additional allocation of the environmental impacts to each one of the functions. The operation of the ships was considered as comprised by 4034 single crossings per year with an average distance of 7 km per crossing and an average service speed of 9.5 knots corresponding with the service currently performed between Kirkwall and Shapinsay. Therefore the ship crosses approximately 28,238 km/year and 847,140 km during the considered lifetime of 30 years.

Impact Assessment Method and Allocations Procedures

The hierarchist perspective of the impact assessment method ReCiPe 2016, hereafter ReCiPe 2016 (H), was selected based on the broad impact categories covered by this method, its global scope and the possibility of considering both mid-point and end-point impacts. The hierarchist perspective is the consensus model most commonly used in scientific models (Huijbregts et al. 2016). The cut-off system model was used as underlying philosophy for the systems taken from the database ecoinvent 3.5.

2.2.3 *Inventory Analysis*

Previous studies have pointed out that the complete life cycle of vehicles involves the life cycle of the fuel, usually called well-to-tank life cycle, and the life cycle of the equipment (Nordelöf et al. 2014). Beyond that, some authors include the infrastructure life cycle in the case of future electric vehicles, i.e. charging systems, to describe the complete life cycle (Marmiroli et al. 2019). Although this approach has been mainly used for vehicles, it can be extended to ships, as is the case in the project HySeas III.

This approach is shown in Fig. 2.1, where the different components considered for ship manufacturing as well as for ship energy supply are displayed for the three different alternatives considered in this work. On one side, the ship is manufactured using different components to accomplish the functions previously described in Table 2.1. On the other side, the energy supply of the ship consists of diesel, electricity or hydrogen, which in this case was considered as produced from wind energy. These two life cycles merge in the use phase of the ship. As mentioned before, the use phase is followed by a final disposal phase, which was not considered here.

Life Cycle of the Ship

Today most of the ships with similar features use marine diesel as fuel or, in the case of bigger ships with low-speed engines, heavy fuel oil. However, diesel was considered in this analysis, since it is the fuel mainly used for small ships. Diesel is burned in internal combustion engines to obtain mechanical energy for propulsion and at the same time heat and electricity for different applications of the ship, usually

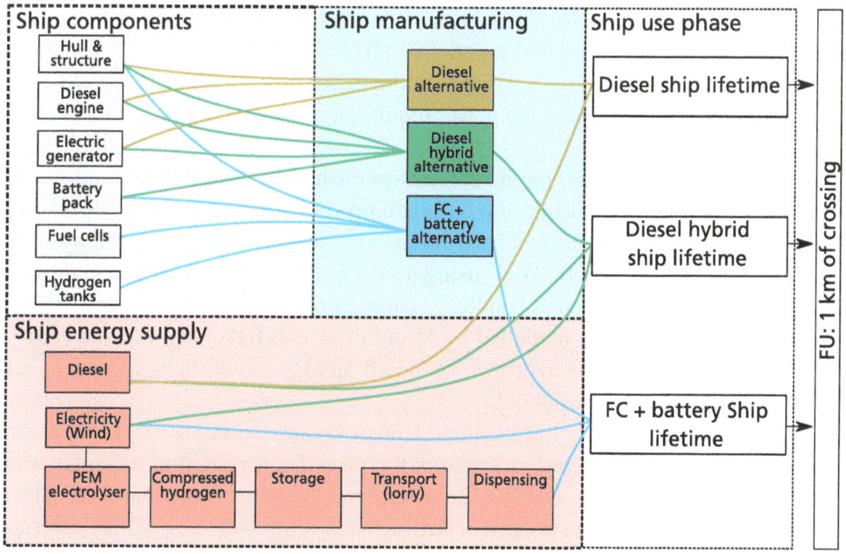

Fig. 2.1 Abstraction of ship and energy supply life cycle as modelled in this work

known as hotel load. In some cases, this is undertaken by a diesel-electric system in which all the mechanical energy is converted to electricity, which is further used for electric thrusters and on-board systems in opposition to the case in which the engine is coupled directly or via a gearbox to a propeller. In comparison to diesel powered systems, fuel cell outputs are heat and electricity. The latter can be used for feeding electric thrusters and any on-board systems in a similar way as done in the case of a diesel-electric system. Additionally, ships can be propelled by a hybrid system of a combustion engine and batteries installed on-board. This is done to operate internal combustion engines at their most efficient point, while batteries can give support during peak conditions by discharging and charging again at low load conditions or from the mains. Batteries produce mainly electricity and therefore are employed mainly in electric propulsion systems.

For purposes of this work, three alternatives differing in terms of their propulsion system and fuels were considered. A first alternative, representing the conventional diesel-electric system, is referred hereafter as diesel electric ship (DES). A second design, considering a diesel electric system assisted by on-board batteries that can be charged by on-board electricity generation or connection to the mains, is named diesel battery electric ship (DBES). Finally, an alternative design using hydrogen stored in high pressure tanks, fuel cells and batteries, as designed in project HySeas III, is referred as fuel cell battery electric ship (FCBES). Since fuel cells and batteries produce electricity, an electric generator was not considered necessary in this case.

The different assumptions considered in this work are summarised in Table 2.2. The diesel engine was modelled by upscaling the inventory in ecoinvent 3.5 for a marine engine construction, which is given in terms of 1000 kg. A 375 kW engine has a mass of approximately 1800 kg (Volvo Penta 2019). Two of those engines would add around 3600 kg, giving a mass scaling factor of 3.6 compared to the system in ecoinvent 3.5. On the other hand, ecoinvent 3.5 includes systems for 200 kW electric generators with a weight of 850 kg. A 500 HP (375 kW) motor has a weight of approximately 1177 kg, giving a weight scaling factor of 1.38. When two generators are considered in order to meet the power requirements, a total scaling factor of 2.77 is obtained.

In the case of fuel cells, the inventories published by Miotti et al. and Bekel and Pauliuk were adapted using ship on-board power (Miotti et al. 2017; Bekel and Pauliuk 2019). Miotti et al. published inventories for an 85 kW fuel cell system, which were scaled up to 600 kW by using a factor of 7.06. Moreover, the hydrogen storage system was modelled as bundles of carbon fibre tanks each containing 5.6 kg of hydrogen, as previously modelled by Miotti et al. and Bekel and Pauliuk (2017, 2019). Given that the total hydrogen on-board storage would have a capacity of 600 kg, 108 tanks were considered.

On-board Lithium-Nickel-Manganese-Cobalt-Oxide (NMC) 1:1:1 lithium ion batteries were also considered as assisting the propulsion system in a similar way as previously stated for the DBES. The inventories for this component were taken as described by Ellingsen et al., and scaled-up according to the size of the batteries considered for the future ship (Ellingsen et al. 2013). Ellingsen et al. described power packs of 22 kWh, which were scaled up to 768 kWh by using a factor of 28.8. The

Table 2.2 Assumptions for ship components according to project specifications

Ship Component	Specification	Lifetime	Reference
Hull and Structure	190 tonnes of steel 20 tonnes of aluminium	30 years	HySeas III
Diesel Engine	2 × 375 kW 40% efficiency	30 years	Scaled from ecoinvent 3.5
Electric generator	2 × 300 kW 99% efficiency	30 years	Scaled from ecoinvent 3.5
Battery set	768 kWh NMC 1:1:1 Li-ion batteries 90% charging efficiency Produced in Germany	10 years (3 battery sets in total during life time)	Ellingsen et al. (2013)
Fuel cells	Proton exchange membrane 600 kW 50% efficiency Pt load: 0.4 mg/cm^2 Lifetime of 20,000 h	7 years (5 fuel cell system changes in total during life time)	Miotti et al. (2017) Bekel and Pauliuk (2019)
Hydrogen tanks	Carbon fibre 600 kg of hydrogen storage 350 bar	30 years	Miotti et al. (2017) Bekel and Pauliuk (2019)

amount of energy assumed for the manufacturing of battery cells was 280 kWh/kg of cells, according to the average value shown by Ellingsen et al. (2013). The German grid contained in ecoinvent 3.5 was assumed for the production of battery cells.

Additionally, the materials for the hull and structure, namely steel and aluminium, were considered similar in all the cases and included as a part of the inventories. The estimations for the amount of materials were provided by Ferguson Marine Ltd and were specified as 190 tonnes of low alloyed hot rolled steel and 20 tonnes of aluminium alloy, metal matrix composite. However, no manufacturing processes were considered for the construction of the ship, because at this stage it is still unknown how much energy and consumables will be required for the construction. Other materials such as glass, furniture or the ship's electronic system were not considered in the analysis because they do not differ considerably between the different alternatives and belong to other functions of the ship.

Ship's Energy Supply

With a view to the ship's energy supply, a lower heating value (LHV) of 42.7 MJ/kg for diesel and its production by the global supply of fossil fuels as modelled in the database ecoinvent 3.5 were assumed. The emissions of its combustion were modelled using a fishing vessel included in the database ecoinvent 3.5. Furthermore, it was assumed that the electricity supply for the hydrogen production originated from

Table 2.3 Energy carriers and electricity consumption of each of the considered ship alternatives

Type of ship	Diesel (kg/crossing)	Electricity (kWh/crossing)	Hydrogen (kg/crossing)
Diesel	54.9	–	–
Diesel hybrid	48.2	40.7	–
FC + battery ship	–	40.7	13.59

Source estimations of HySeas III consortium

wind power, as it is the current situation in the Orkney Islands, where the future prototype will operate. Consequently, hydrogen was considered as produced using wind electricity and a proton exchange membrane (PEM) electrolyser to obtain compressed hydrogen. The electricity consumption for hydrogen production was assumed to be 50 kWh/kg H_2 and the inventories for PEM electrolysers were modelled according to the inventories published by Wulf and Kaltschmidt (2018). After being produced and compressed, hydrogen is stored in trailers with a capacity of 200 kg at 350 bar, which are then conveyed for a distance of 6 km. For calculation purposes a distance of 12 km was considered, as the trailers must be driven back to the production site. Finally, the energy carrier is delivered to the ship by a dispensing unit, which in this case was assumed to be similar to a natural gas dispensing unit, in the same line as the assumption done by Wulf and Kaltschmidt (2013).

The energy consumption was modelled according to the current calculations performed within project HySeas III and considering the different efficiencies mentioned in Table 2.2. The consumption of the different energy carriers and electricity per crossing is shown in Table 2.3. According to these assumptions, the diesel prototype would have a yearly fuel consumption of 221,579 kg or 263,784 l when a density of 0.840 kg/l is considered for this fuel. Current consumption of MF Shapinsay, the ferry serving this route at present time was 170,400 l in 2018 (Orkney Island Council. Passengers, vehicles and fuel consumption of Orkney Marine Services Ships (Personal communication), 20/03/2019). Although the MF Shapinsay is smaller than the prototype developed in project HySeas III, making comparisons difficult, these figures give an idea of the magnitude of fuel consumption.

2.3 Results: Impact Assessment and Interpretation

The following section describes the results obtained for the midpoint and endpoint assessment and the single scores obtained from the ReCiPe 2016 (H) method.

2.3.1 Midpoint Characterisation

The main motivation for using hydrogen as fuel and fuel cells as energy converter lies in the reduction of GWP emissions. As summarized in Table 2.4, when wind power and PEM electrolysis are used to produce hydrogen which supplies FCBES propulsion, the reduction of GWP from cradle to end-of-use is approximately 89% in comparison to the DES alternative. The DBES alternative together with electricity produced using wind power may allow reductions of approximately 8% compared to the reference DES case. FCBES has additional lower impacts regarding SOD, IR, OFHH, FPMF, OFTE, TAC and FRS. On the other hand, the FCBES shows higher impacts compared to the DES and DBES in terms of FEU, MEU, TEC, FEC, MEC, HCT, HNCT, LU, MRS and WATC. The results for the DBES in comparison with the DES alternative are 6% higher in SOD and 25% higher in IR.

A comparison between the different alternatives is shown in Figs. 2.2 and 2.3 by normalising to the highest value obtained in each impact category. Additionally, the contributions of each ship component and energy supply involved in the life cycle are displayed. Most of the impact in the categories GWP, SOD, IR, OFHH, FPMF, OFTE and FRS in the cases of the DES and DBES alternatives as displayed in are due to the use of diesel for propulsion, mainly due to the tailpipe emissions, which are avoided in the operation with hydrogen and electricity.

Most of the effects in which the FCBES alternative has higher impact, as displayed in Fig. 2.3, are derived from the mining and refining processes of the materials employed for the ship's batteries and fuel cells as well as for wind turbines and electrolysers for hydrogen production. Therefore, these impacts are not located where the

Table 2.4 Characterisation results of the different three alternatives using ReCiPe 2016 (H). The lowest impact is highlighted in green, middle impact in yellow and highest impact in red

Impact category	Unit	DES	DBES	FCBES
Global warming potential (GWP)	kg CO_2 eq/km	2.96×10^{1}	2.70×10^{1}	3.31×10^{0}
Stratospheric ozone depletion (SOD)	kg CFC11 eq/km	7.24×10^{-6}	7.27×10^{-6}	2.15×10^{-6}
Ionizing radiation (IR)	kBq Co^{-60} eq/km	3.51×10^{-1}	4.42×10^{-1}	2.63×10^{-1}
Ozone formation, Human health (OFHH)	kg NO_x eq/km	6.37×10^{-1}	5.63×10^{-1}	9.48×10^{-3}
Fine particulate matter formation (FPMF)	kg $PM_{2.5}$ eq/km	2.06×10^{-1}	1.85×10^{-1}	1.04×10^{-2}
Ozone formation, Terrestrial ecosystems (OFTE)	kg NO_x eq/km	6.40×10^{-1}	5.66×10^{-1}	9.82×10^{-3}
Terrestrial acidification (TAC)	kg SO_2 eq/km	6.52×10^{-1}	5.85×10^{-1}	2.61×10^{-2}
Freshwater eutrophication (FEU)	kg P eq/km	9.88×10^{-4}	2.54×10^{-3}	3.48×10^{-3}
Marine eutrophication (MEU)	kg N eq/km	8.45×10^{-5}	1.80×10^{-4}	3.14×10^{-4}
Terrestrial ecotoxicity (TEC)	kg 1,4-DCB/km	2.21×10^{1}	4.09×10^{1}	4.58×10^{1}
Freshwater ecotoxicity (FEC)	kg 1,4-DCB/km	1.07×10^{-1}	3.05×10^{-1}	9.09×10^{-1}
Marine ecotoxicity (MEC)	kg 1,4-DCB/km	1.75×10^{-1}	4.50×10^{-1}	1.19×10^{0}
Human carcinogenic toxicity (HCT)	kg 1,4-DCB/km	4.76×10^{-1}	5.72×10^{-1}	1.07×10^{0}
Human non-carcinogenic toxicity (HNCT)	kg 1,4-DCB/km	3.40×10^{0}	9.50×10^{0}	1.45×10^{1}
Land use (LU)	m^2a crop eq/km	6.05×10^{-2}	9.42×10^{-2}	1.77×10^{-1}
Mineral resource scarcity (MRS)	kg Cu eq/km	3.14×10^{-2}	2.28×10^{-1}	2.97×10^{-1}
Fossil resource scarcity (FRS)	kg oil eq/km	9.83×10^{0}	8.89×10^{0}	8.80×10^{-1}
Water consumption (WATC)	m^3/km	5.20×10^{-2}	6.68×10^{-2}	8.55×10^{-2}

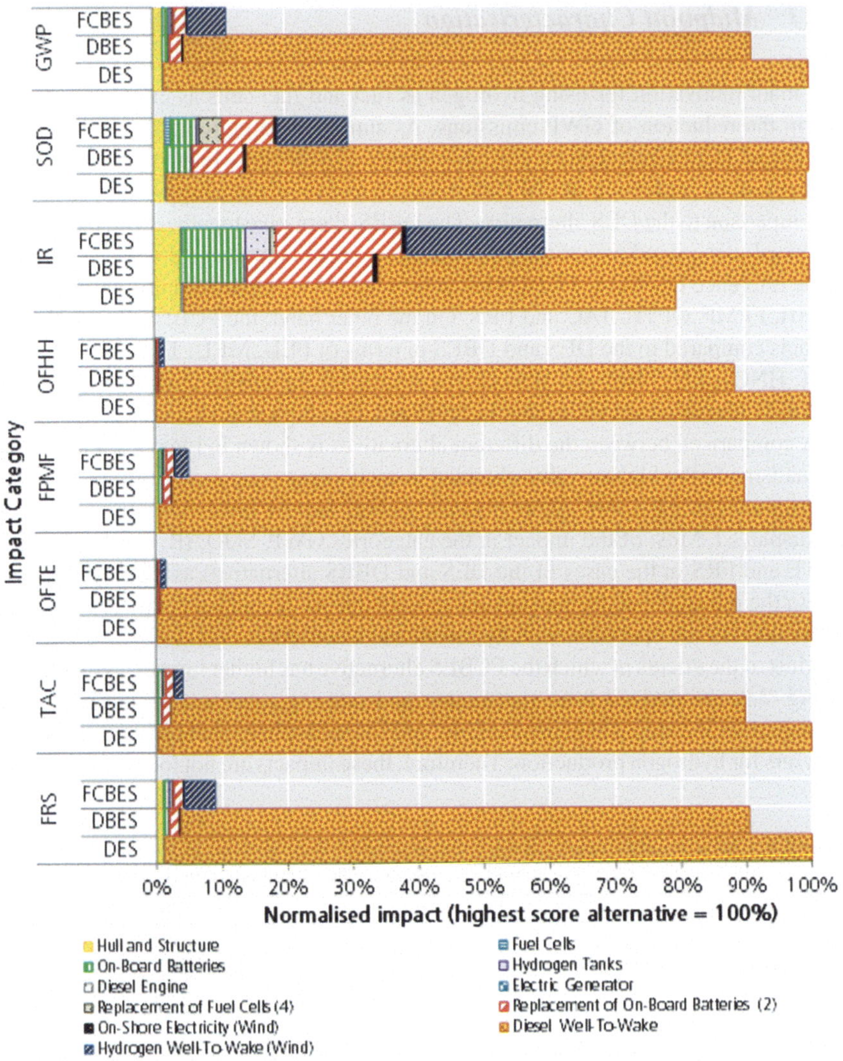

Fig. 2.2 Comparison of the midpoint impact assessment results for diesel electric ship (DES), diesel battery electric ship (DBES) and fuel cell battery electric ship (FCBES) RoPax ferry alternatives in the categories in which the FCBES has lower impact. Results normalised to the highest total impact alternative in each of the categories

ship operates, but where the materials used in the manufacture of different components are sourced. For FCBES, both fuel cells and battery replacements gather an important share of the impact, particularly in the impact categories SOD and IR.

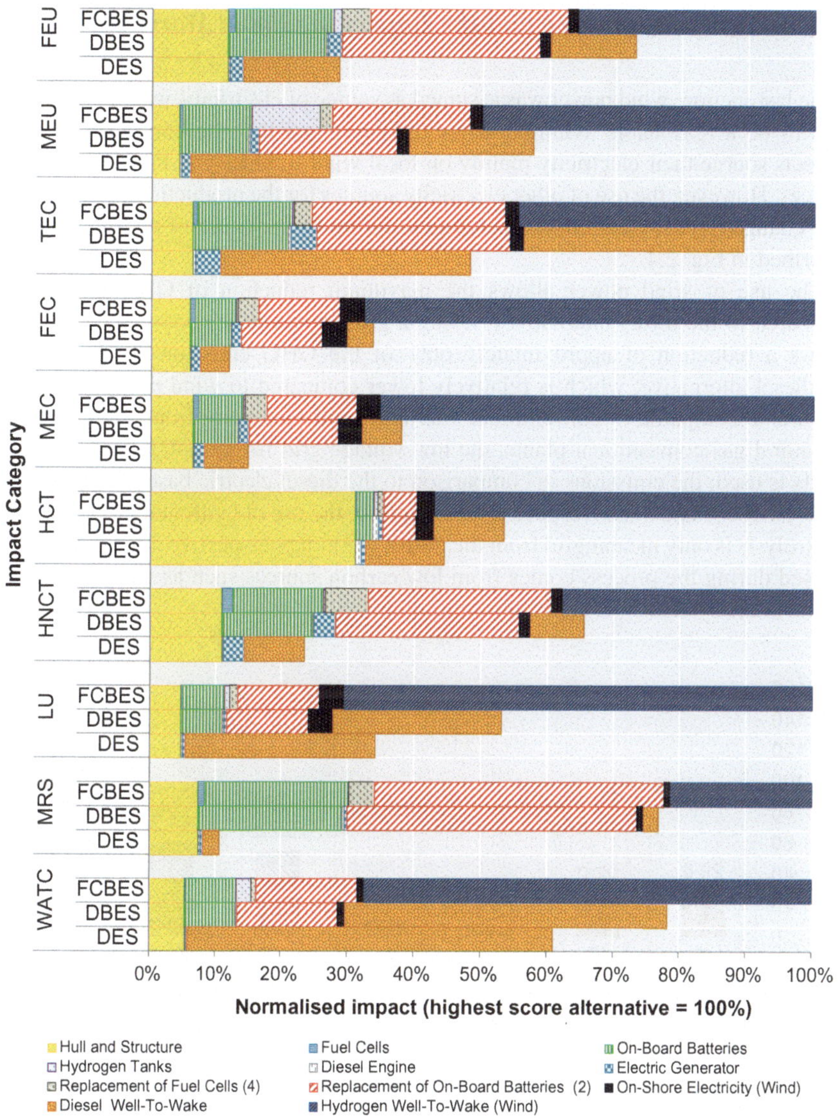

Fig. 2.3 Comparison of the midpoint impact assessment results for diesel electric ship (DES), diesel battery electric ship (DBES) and fuel cell battery electric ship (FCBES) RoPax ferry alternatives in the categories in which the FCBES has higher impact. Results normalised to the highest total impact alternative in each of the categories

2.3.2 Electricity Source and the Impact in Global Warming

In the last section, wind power was assumed as source of electricity, which is considered to be a reasonable assumption for Orkney given that the current hydrogen projects source their electricity mainly on local wind turbines and tidal conversion devices. However, the use of other electricity sources for the production of hydrogen may change radically the GHG emissions resulting from the use of hydrogen, as is described in Fig. 2.4.

The use of wind power allows the maximum reduction of GHG emissions compared to the diesel alternatives. Using a ground mounted photovoltaic system allows a reduction of approximately 66% of the GHG emissions compared to the diesel alternative, which is relatively lower compared to wind power, but still favourable compared to conventional alternatives. If electricity from sources such as natural gas conventional plants, the low voltage grid from the UK or oil power plants is used, the emissions in comparison to the diesel electric base case increase by 86%, 103% and 366% respectively. Therefore, the use of hydrogen produced via electrolysis is only meaningful from the global warming perspective if the electricity utilised during the process comes from low carbon sources such as wind power or photovoltaic systems.

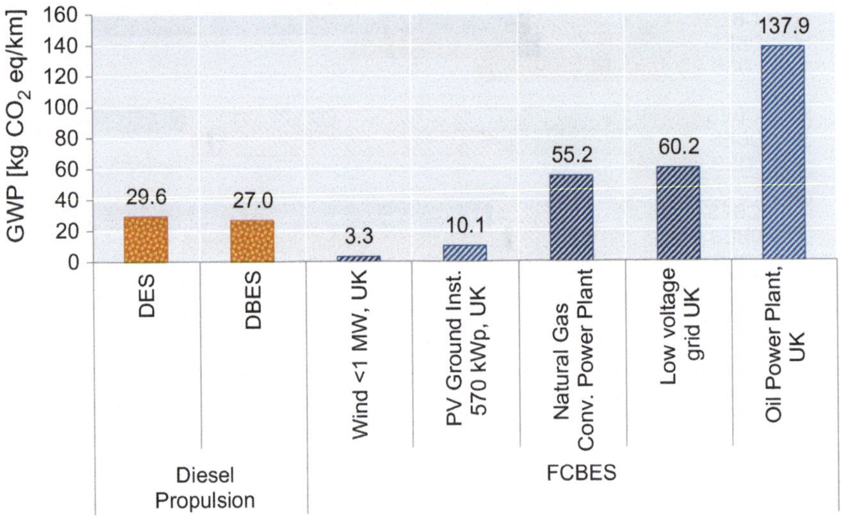

Fig. 2.4 Comparison of Global Warming Potential (GWP) for diesel electric ship (DES), diesel battery electric ship (DBES) and fuel cell battery electric ship (FCBES) employing different electricity sources for hydrogen electrolysis

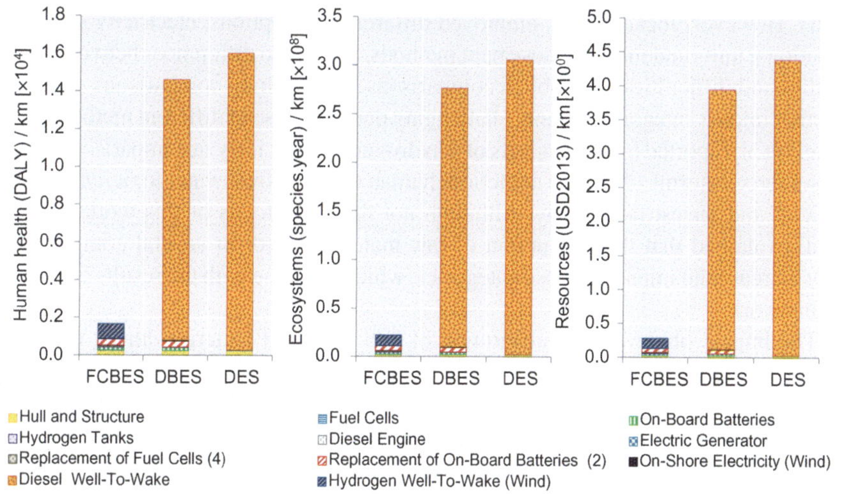

Fig. 2.5 Comparison of the endpoint assessment results for diesel electric ship (DES), diesel battery electric ship (DBES) and fuel cell battery electric ship (FCBES) RoPax ferry alternatives. From left to right: damage to human health ($\times 10^4$), to ecosystems ($\times 10^8$) and to resource availability ($\times 10^0$)

2.3.3 Endpoint Characterisation

The ReCiPe 2016 method allows endpoint characterisation by assigning the impacts to three endpoint categories related to protection areas. These endpoint categories are known as damage to human health, damage to ecosystems and damage to resource availability. The indicators for the impacts are reported in terms of Disability-Adjusted Life Year (DALY), species/year and increased cost respectively.

Figure 2.5 shows that the FCBES alternative has the lowest impact in all the three impact categories, followed by DBES and DES. This suggests that although the FCBES has higher midpoint impacts in some categories, the overall impact among the protection areas is smaller. Most of the endpoint impacts of DES and DBES alternatives are generated by the combustion of diesel in terms of impacts to the human health and ecosystems and the consumption of this fossil fuel in terms of resources.

2.4 Discussion

Jokela et al. described a GWP reduction of 79% of a hydrogen electric ferry compared with a diesel ferry (Jokela et al. 2019). This is 10% lower than the one found in this study. Moreover, the work done by Jokela et al. described a GWP of 5.2 kg CO_2eq for the 3 km crossing. If this amount is adjusted to 1 km of crossing, an amount of 1.7 kg CO_2eq/km would be obtained, which is 48% lower than the one obtained in this study. However, Jockela et al. employed different assumptions, electricity sources, functional units and impact assessment methods, leading to differences between both studies and difficulties to establish comparisons. Establishing comparisons in terms of other impact categories is more challenging due to the use of different methods, but Jokela et al. reported higher impacts of a hydrogen electric ferry in comparison with a diesel ferry in terms of abiotic depletion, human toxicity, freshwater toxicity, marine toxicity and terrestrial toxicity. Although not the main focus of this work, Jokela et al. explained that the production of raw materials used for several components may increase the impact in those categories, which is in line with the results obtained in this work.

The impact observed by substitution of fuel cells and batteries during the lifetime was considerable. Therefore, maximising the lifetime of these components may contribute to diminishing the impact in most of the impact categories. Future improvements in the manufacturing of batteries or fuel cells were not considered. These improvements may be for instance a reduction in the amount of platinum used in fuel cells, different cell chemistries for Li-ion batteries or a reduction of the energy used for producing these elements. Moreover, proper recycling may allow the reduction of the impact for obtaining new materials and therefore diminish the impact on some of the categories, but this was not considered in the present work.

The analysis using different electricity sources showed that wind power allows the maximum reduction of GWP among the considered alternatives. Other low carbon sources such as hydropower or nuclear power may also be used to produce hydrogen, but may not be available at some locations, and may face additional problems for their implementation.

The endpoint analysis showed that the FCBES has the lowest impact to human health, environment and resources, even when the impact in the midpoint categories is higher compared to the DBES and DES alternatives. Despite the fact that endpoint categories have more policy relevance because their indicators are less abstract, it is also known that they introduce more uncertainty in the results.

2.5 Conclusions

The results obtained in this study show that a hydrogen fuel cell and battery electric ferry may allow the reduction of global warming impacts from cradle to end-of-use of up to 89% when compared with a conventional diesel electric ferry on the route Kirkwall–Shapinsay. However, if hydrogen is obtained using electrolysis, low carbon electricity sources such as wind power or photovoltaic modules should be used in order to maintain the benefits related to the reduction of greenhouse gas emissions.

The use of hydrogen and fuel cells in conjunction with batteries also decreases the potential life cycle impact in terms of stratospheric ozone depletion, ionizing radiation, ozone formation–human health, fine particulate matter formation, ozone formation–terrestrial ecosystems, terrestrial acidification and fossil resource scarcity. The impacts in other categories such as fresh water eutrophication, marine eutrophication, terrestrial ecotoxicity, freshwater ecotoxicity, marine ecotoxicity, human carcinogenic toxicity, human non-carcinogenic toxicity, land use, mineral resource scarcity and water consumption were higher in comparison with the diesel and diesel battery alternatives. The main reason for this lies in the sourcing of the materials employed in the manufacturing of the components used to build the hydrogen powered fuel cell or in the hydrogen supply chain. The components considered for the hydrogen powered fuel cell and battery ferry include batteries, fuel cells and high pressure tanks. Hydrogen production includes the use of wind turbines for producing electricity and electrolysers to convert electricity to chemical energy stored in hydrogen, among others. Thus, these impacts are not generated locally in the place where the ship is operating, but in the places from which the materials used for the construction of the ship or wind turbines are obtained. The reuse of some of this components or recycling of their materials may close the loop and diminish the environmental impact of obtaining these materials. However, this was not explored in this work.

On the other hand, a diesel battery electric ship allows a reduction of global warming potential of 8% compared to a conventional diesel electric ferry when recharged using wind power. However, an increase of 6% in stratospheric ozone depletion and 25% in ionizing radiation was observed due to the production processes of batteries and their replacement along the lifetime of the ship.

Particularly, the replacement of ship components to accomplish the considered 30 years lifetime has a relevant impact in most of these categories. Surprisingly, batteries had a higher impact in relation to fuel cells in the case of the hydrogen powered fuel cell battery ship. In general, the extension of the lifetime of both fuel cells and batteries must be an important goal to minimise the environmental impact along the lifetime.

The endpoint analysis showed that the hydrogen powered fuel cell battery ship has a lower impact in terms of damage to the human health, damage to the ecosystem and damage to resource availability compared to the diesel electric and diesel battery electric alternatives, even when the impact for some of the midpoint categories are higher.

2.6 Future Work

The results presented in this study are related to the analysis of the concept developed in the first stages of the project HySeas III. The future power string trials and ship construction will allow access to specific data that can increase the level of detail and the quality of the analysis. Further comparisons with other ship propulsion architectures will be done in the near future. Additionally, other economic and social analysis will be conducted during the project to have a holistic perspective of this new application.

Acknowledgements HySeas III is led by the shipyard Ferguson Marine (Scotland) and the University of St Andrews (Scotland). Other partners are Orkney Islands Council (Scotland), Kongsberg Maritime (Norway), Ballard Power Systems Europe (Denmark), McPhy (France), Interferry (Belgium/US) and DLR–German Aerospace Center acting as main contributor to this study. This project has received funding from the European Union's Horizon 2020 research and innovation programme under grant agreement No 769417. We would also like to thank all project partners of HySeas III who kindly contributed with data and feedback. Finally, we would like to thank Danielle Schmidt (DLR Institute of Networked Energy Systems) who proof-read the writing style.

References

Bekel K, Pauliuk S (2019) Prospective cost and environmental impact assessment of battery and fuel cell electric vehicles in Germany. Int J Life Cycle Assessment 24, 2230–2237

BIG HIT Project (2019) About the project. https://www.bighit.eu/. Last access 02/10/2019

DIN Deutsches Institut für Normung e.V., Umweltmanagement – Ökobilanz – Anforderungen und Anleitungen (ISO 14044:2006); Deutsche und Englische Fassung EN ISO 14044:2006, Berlin, 2006

DIN Deutsches Institut für Normung e.V., Umweltmanagement – Ökobilanz – Grundsätze und Rahmenbedingungen (ISO 14040:2006); Deutsche und Englische Fassung EN ISO 14040:2006, Berlin, 2006

DNV GL–Maritime, Maritime Forecast to 2050—Energy Transition Outlook 2019. https://eto.dnvgl.com/2018/maritime. Last accessed 02/10/2019

Ellingsen LAW, Majeau-Bettez G, Singh B, Srivastava AK, Valøen LO, Strømman AH (2013) Life cycle assessment of a lithium-ion battery vehicle pack. J Industrial Ecol 18(1)

European Commission (2019) Reducing emissions from the shipping sector. https://ec.europa.eu/clima/policies/transport/shipping_en#tab-0-0. Last access 02/10/2019

Gilbert P, Walsh C, Traut M, Kesieme U, Pazouki K, Murphy A (2018) Assessment of full life-cycle air emissions of alternative shipping fuels. J Cleaner Prod 855–866

Huijbregts M, Steinmann Z, Elshout P, Stam G, Verones F, Vieira M, Hollander A, Zipj M, van Zelm M ReCiPe 2016 v1.1—A harmonized life cycle impact assessment method at midpoint and endpoint level—Report I: Characterization—RIVM Report 2016-0104a, 2017. https://www.rivm.nl/documenten/recipe2016v11. Last access: 07/10/2019

HySeas III Project (2019) The Project. https://www.hyseas3.eu/the-project/. Last access 02/10/2019

IHS Markit (2019) SeaWeb. https://maritime.ihs.com/Account2/Index. Last access 02/10/2019

International Maritime Organization (2018) Annex 11—Resolution MEPC.304(72)—Initial IMO strategy on reduction of GHG emissions from ships. 13/04/2018. http://www.imo.org/en/OurWork/Documents/Resolution%20MEPC.304(72)%20on%20Initial%20IMO%20Strategy%20on%20reduction%20of%20GHG%20emissions%20from%20ships.pdf. Last access 02/10/2019

Jokela AM, Polden IG, Tanum OS, Seim V (2019) Livsløpsanalyse av hydrogen-elektrisk ferje. 28/05/2018. https://hvlopen.brage.unit.no/hvlopen-xmlui/handle/11250/2573591. Last access 02/10/2019

Marmiroli B, Dotelli G, Spessa E (2019) Life cycle assessment of an on-road dynamic charging infrastructure. Appl Sci, Aug 2019, 2117

Miotti M, Hofer J, Bauer C (2017) Integrated environmental and economic assessment of current and future fuel cell vehicles. Int J Life Cycle Assessment 22:94–110

Nordelöf A, Messagie M, Tillman AM, Söderman ML, Van Mierlo J (2014) Environmental impacts of hybrid, plug-in hybrid, and battery electric vehicles—what can we learn from life cycle assessment? Int J Life Cycle Assess 1866–1890

Orkney Island Council. Passengers, vehicles and fuel consumption of Orkney Marine Services Ships (Personal communication), 20/03/2019

Papanikolaou A (2019) Holistic ship design optimisation. In: A holistic approach to ship design—Volume 1: optimisation of ship design and operation for life cycle. Springer Nature, Cham, Switzerland AG, 2019, pp 9–42

Point and Sandwick Trust (2019) Scottish Western Isles Ferry Transport using Hydrogen (SWIFTH2)—Feasibility Report. 12/05/2019. http://www.pointandsandwick.co.uk/wp-content/uploads/2019/07/Scottish-Western-Isles-Ferry-Transport-using-Hydrogen-Feasibility-Report.pdf. Last access 02/10/2019

Smith T, O'Keeffe E, Aldous L, Parker S, Raucci C, Traut M (2015) Third IMO GHG Study 2014. http://www.imo.org/en/OurWork/Environment/PollutionPrevention/AirPollution/Documents/Third%20Greenhouse%20Gas%20Study/GHG3%20Executive%20Summary%20and%20Report.pdf. Last accessed 02/10/2019

Surf 'N' Turf Project (2019) The Surf 'N' Turf Project. http://www.surfnturf.org.uk/. Last access 02/10/2019

Volvo Penta, "IPS650/800" (2019) https://pubs.volvopenta.com/ProdDocs/Home/Disclaimer?publication=47709257&lang=en-US. Last access: 08/10/2019

Wulf C, Kaltschmidt M (2018) Hydrogen supply chains for mobility—environmental and economic assessment. Sustainability 10:1699

Wernet G, Bauer C, Steubing B, Reinhard J, Moreno-Ruiz E, Weidema Bo (2016) The ecoinvent database version 3 (part I): overview and methodology. Int J Life Cycle Assessment 21(9):1218–1230

Chapter 3
Analysis of Fuel and Powertrain Combinations for Heavy-Duty Vehicles from a Well-to-Wheels Perspective: Model Development and Sample Application

Mara Kuttler and Simon Pichlmaier

Abstract This article applies the Well-to-Wheels (WTW) methodology in order to evaluate to which extent alternative fuels and powertrains can contribute to GHG emissions reductions within the heavy-duty segment of the European Union (EU). The analysis compares the WTW chain of diesel as the fossil benchmark to biodiesel, bio-methane, Power-to-Liquid (PtL) and Power-to-Gas (PtG) fuels, as well as catenary electric vehicles (CEVs) in heavy-duty tractor-trailers. The WTW analysis found that fuel and powertrain combinations from renewable energy sources offer a high potential for reducing GHG emissions as compared to diesel, however, the magnitude of the reductions significantly varies depending on feedstock and fuel type. Taking feedstock availability into account, the direct use of electricity in CEVs was identified as the most promising option for decarbonizing the heavy-duty long-haul sector, although the performance of CEVs depends on the extent to which the required energy comes from non-fossil resources.

Keywords Well-to-Wheels analysis · Alternative fuels · Heavy-duty vehicles

3.1 Introduction

Heavy-duty vehicles (HDVs) contribute disproportionally to greenhouse gas (GHG) emissions within the transport sector of the European Union (EU). Even though they just represent 5% of the EU vehicle fleet, they cause one fifth of the GHG emissions in 2016 from all means of transport[1] and 27% of the emissions from road transport (Carbon Dioxide Emissions from Europe's Heavy-Duty Vehicles 2018; Too Big to Ignore — Truck CO_2 Emissions in 2030 2015). Thus, decarbonizing the

[1]Including international aviation, but excluding international shipping.

M. Kuttler (✉) · S. Pichlmaier
Forschungsstelle Für Energiewirtschaft e.V., Am Blütenanger 71, 80995 Munich, Germany
e-mail: mkuttler@ffe.de

S. Albrecht et al. (eds.), *Progress in Life Cycle Assessment 2019*,
Sustainable Production, Life Cycle Engineering and Management,
https://doi.org/10.1007/978-3-030-50519-6_3

heavy-duty portion of the transport sector is a priority in order to reach the EU-wide climate targets, especially since the demand for road freight transport is continually increasing. Means to reduce GHG emissions and to transition away from crude oil as the dominant energy source would be to adopt alternative fuels and powertrains. To assess the extent that these options contribute to GHG emissions reductions, Well-to-Wheels (WTW) analysis is currently the prevailing methodology, especially at the regulatory level (e.g. (Directive (EU) 2018)). Since the 1990s, a number of WTW studies on passenger cars have been carried out (Alamia 2016) and in recent years, the scope of WTW research increasingly expanded to the heavy-duty segment. However, since the results of these studies are only valid in their respective contexts, their comparability is limited, what makes data required for decision making hard to assess in specific situations. This article presents the development of a model that allows a WTW comparison between the fuel and powertrain combinations that are currently considered to replace the combustion of diesel in the heavy-duty segment of the EU. Moreover, a sample analysis is carried out with the model that assesses the emissions savings potential of using selected alternative fuels as compared with using diesel. As regards catenary electric vehicles (CEVs), there is a special focus on modeling the carbon intensity of the electricity mixes of different EU countries.

3.1.1 The Application of Alternative Fuels in HDVs

The European Commission (EC) defines HDVs as "freight vehicles of more than 3.5 tons (trucks) or passenger transport vehicles of more than 8 seats (buses and coaches)" (Questions and Answers on the Commission Strategy for Reducing Heavy-Duty Vehicles' (HDVs) Fuel Consumption and CO_2 Emissions [Press Release] 2014). Further distinctions can be made according to weight class, mission profile (urban, regional, long haul) and vehicle configuration (e.g. tractor-trailer) [Commission Regulation (EU) 2017 Annex I]. Depending on these characteristics, different alternative fuels, "which serve, at least partly, as a substitute for fossil oil sources in the energy supply to transport […]" are currently discussed (Directive 2014 Article 2). As illustrated by Fig. 3.1, the landscape of alternative fuels and powertrains applicable to the HDV segment is broad and most fuel types can be generated from several energy sources, including both fossil and regenerative options.

The easiest transition towards alternative fuels is the introduction of so-called "drop-in fuels" that are similar to diesel on a molecular level and are fully compatible with existing engine technologies and refueling infrastructure, such as synfuels and biofuels in blends up to a certain share (Neuling 2017; Marelli 2015). An alternative to the use of drop-in fuels is the development of internal combustion engines (ICEs) dedicated to the use of alternative fuels, such as dedicated gas engines. In contrast to the self-ignited diesel cycle engines, where heat is generated by the compression of air, dedicated gas engines use positive ignition (PI) systems (The Future of Trucks — Implications for Energy and the Environment 2017; Schaub et al. 2007). Compression ignition (CI) ICEs generally have higher thermal efficiencies than PI

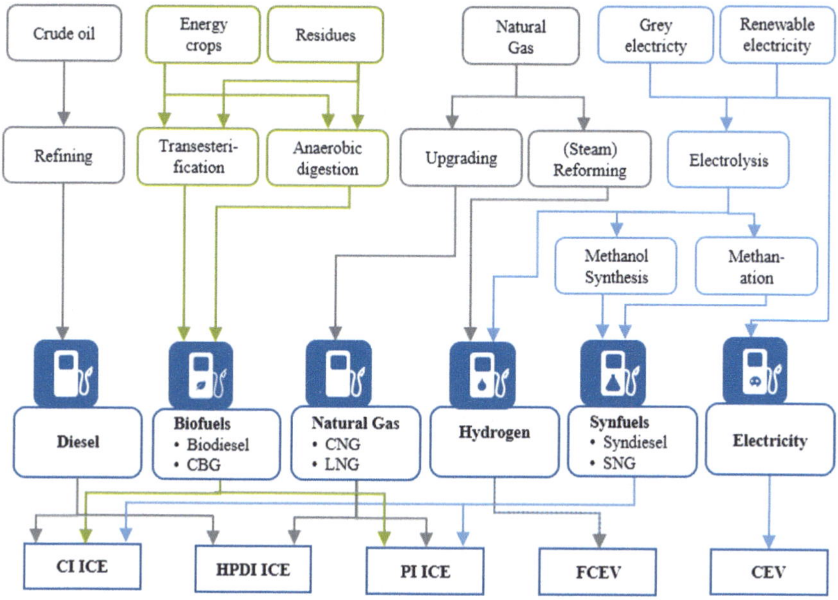

Fig. 3.1 Analyzed fuel and powertrain combinations and key elements of the fuel synthesis process, based on (Edwards 2014)

ICEs (Isenstadt 2017), however using natural gas (NG) as vehicle fuel offers possible GHG savings due to the lower carbon content of methane (CH_4) in comparison to diesel (The Future of Trucks — Implications for Energy and the Environment 2017). High Pressure Direct Injection (HPDI) ICEs, on the other hand, are diesel cycle engines designed to run on liquefied CH_4, along with a small amount of pilot fuel to provoke ignition. However, the HPDI technology offers the advantage that it combines the lower emission factor of NG with the high efficiencies of a diesel engine (Delgado 2015). The application of LNG is better suited for large vehicles that travel over extended driving ranges (annual mileage of 100,000 km or more), since the required in-vehicle storage volumes are lower than those of compressed CH_4 (CNG), although they require still about double the volume than fossil diesel to achieve comparable travel distances (The Future of Trucks — Implications for Energy and the Environment 2017).

Besides combustion engines, technologies that are based on electrified powertrains have been emerging. For the electrification of long-haul operations, CEVs, which are supplied with electricity through catenary overhead lines, have gained special attention since their battery capacities do not have to be as high and holding times for battery charging disappear (dena-Leitstudie Integrierte Energiewende - Zwischenfazit Impulse und Erkenntnisse aus dem 2017). In addition, previous analyses found CEVs to be a more cost-effective option than battery electric vehicles for decarbonizing the heavy-duty sector (Conrad et al. 2019). Another application of electric drivetrains is the Fuel Cell Electric Vehicle (FCEV) that uses hydrogen for the

Table 3.1 Denotation of assessed fuel types

Abbreviation	Fuel and feedstock
B0	Conventional diesel
B_sfl	Biodiesel, sunflower
B_soy	Biodiesel, soy
B_plm	Biodiesel, palm fruit
B_res	Biodiesel, residues (= waste cooking oil)
B_wnd	Syndiesel, wind
CBG_ogw	Compressed Biogas, organic waste
CBG_wmr	Compressed Biogas, wet manure
CBG_mze	Compressed Biogas, maize
SNG_wnd	Synthetic Natural Gas, electricity, wind energy, compressed
SNG_mix	Synthetic Natural Gas, electricity, EU-mix, compressed
CNG	Compressed Natural Gas (= fossil)
LNG	Liquefied Natural Gas (= fossil)
LNG + B0	Liquefied Natural Gas and diesel pilot
H2_NG	Hydrogen, natural gas, compressed
H2_wnd	Hydrogen, electricity, wind energy, compressed
H2_mix	Hydrogen, electricity, EU-mix, compressed
E_wnd	Electricity, wind energy
E_mix	Electricity, EU-mix

on-board generation of electric power. The hydrogen options assessed in this study are generated either via steam reforming of NG, or via electrolysis, the separation of water into hydrogen and oxygen. Besides using hydrogen as final energy carrier in fuel cells, it can be transformed into hydrocarbons of different chain lengths, resulting in the above-mentioned synfuels. This study considers two different synthesis routes for synfuels, first the synthesis of a liquid hydrocarbon (referred to as syndiesel) via the methanol route,[2] and second, the synthesis of a gaseous hydrocarbon (referred to as SNG) via methanation.

The assessed biofuels include fuels from energy crops, which have been highly criticized due to the associated land use change (LUC) implications, as well as biofuels from wastes and residues, such as used cooking oil or manure. The latter are not associated with LUC risks, but diverting wastes and residues from their existing uses might nevertheless be critical and the availability of these feedstocks is limited. According to the International Council on Clean Transportation (ICCT), all the wastes and residues sustainably available in the EU can supply no more than 16% of transport fuels in 2030 (Harrison 2014). Table 3.1 provides an overview of

[2]Using the MtSynfuels process by Lurgi.

the denotations for the different fuels and respective feedstocks that will be used throughout this paper.

3.1.2 WTW Methodology Versus LCA

The WTW methodology evaluates impacts of transport services from the primary energy base (well) to the point where the wheels are turning (wheels) (Kirsten 2016). In contrast to LCA, that quantifies environmental burdens and their potential impacts over the whole life cycle of a product or service, the WTW approach focuses on fuel provision and conversion in the vehicle (Dreier 2018). Consequently, WTW analysis does not consider the manufacturing of vehicles or equipment used for fuel production and distribution, nor does it include any disposal aspects (Edwards 2014). WTW analysis is usually broken down into two parts, one concerned with fuel provision, called the Well-to-Tank (WTT) section, and one concerned with vehicle operation, called the Tank-to-Wheels (TTW) section, with the boundary between the two sections being the vehicle's tank. Figure 3.2 provides an overview of the process steps the two sections typically comprise.

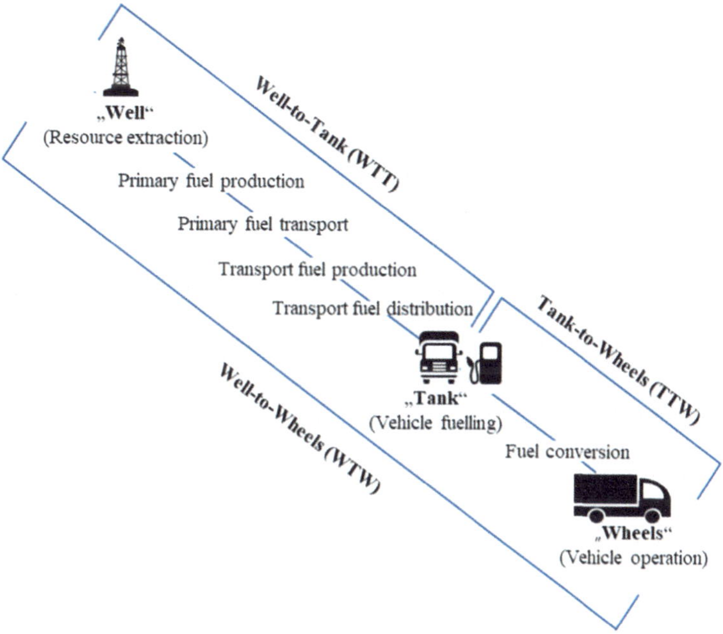

Fig. 3.2 Typical parts of a WTW analysis, based on (Edwards 2014)

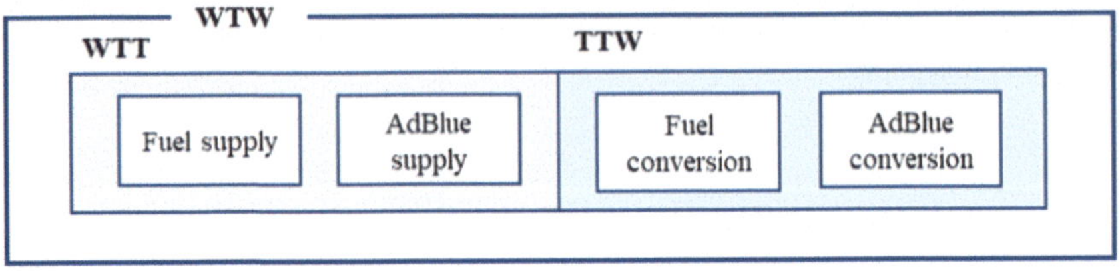

Fig. 3.3 Scope of the WTW model

3.2 Method

This section explains the method that is used to develop a WTW model allowing the comparison between a wide range of fuel and powertrain combinations.

3.2.1 Scope and System Boundaries

The model accounts for fuel and AdBlue provision in the WTT section, and for fuel and AdBlue conversion on-board the vehicle in the TTW section (see Fig. 3.3). The boundary between the two sections is defined as the vehicle's full energy storage. Any losses that occur during vehicle refueling, e.g. electricity losses during battery charging, are thus accounted for by the WTT section.

GHG emissions considered in the model are CO_2, CH_4 and N_2O. CH_4 and N_2O emissions were converted to CO_2-equivalent (CO_2e) using global warming potential (GWP) values as defined in the Fourth IPCC Assessment Report (AR4) for a time horizon of 100 years (GWP_{100}).[3] This approach follows the provisions set out by the United Nations Framework Convention on Climate Change (UNFCCC) reporting guidelines, which define rules for the reporting on national GHG inventories (Report of the Conference of the Parties on its Nineteenth Session, Held in Warsaw from 11 to 23 November 2013). Other GHGs, such as hydrofluorocarbons (HFCs), perfluorocarbon (PFCs) or sulfur hexafluoride (SF6) were not taken into account due to limited data availability. Furthermore, the model does not account for emissions from LUC (direct/indirect) as their magnitude is difficult to quantify and estimations differ significantly across studies. Emissions from LUC are nevertheless important, which is why they are investigated within the sensitivity analysis in Sect. 3.3.

[3]CO_2e factors according to AR4: CO_2:1, CH_4: 25, N_2O: 298 (Pachauri 2007).

3.2.2 General Assumptions

Some general assumptions are made regarding accounting methods for co-products, the application of carbon credits and the efficiency of renewable electricity. To account for co-products generated by processes, such as fuel production, the substitution method was used. According to this method, a credit, which equals the emissions that would have been required to produce the original product, which is replaced by the co-product, is applied. Moreover, biomass feedstock is considered carbon neutral when combusted as the CO_2 emitted during combustion was absorbed from the atmosphere during the growth of the plant. To account for the timely-closed carbon cycle, a credit for biogenic carbon was applied at the WTT side. The same approach was used for CO_2 absorbed in the synthesis process of synfuels (e.g. syndiesel, SNG), which corresponds to the amount of CO_2 emitted during the combustion process in the vehicle. The underlying assumption is that the supplied carbon had already been accounted for in the (industrial) process that generated the CO_2. The provision of renewable electricity was assumed to come from an unlimited energy source and does not consider any regulatory, market-based or indirect technical effects. Different from LCA, electricity supply from renewable energy sources (RES) was thus assumed to be provided with an efficiency of 100%[4] and not to generate any GHG emissions, which is in line with the JEC WTT Report[5] and the default values indicated by the IPCC (Koffi 2017). The conversion of electricity in the vehicle (TTW section) was assumed not to emit any GHG as well as specified in DIN EN 16258.

3.2.3 Well-to-Wheels Data

The calculations of the emissions balances are based on the methodology provided by DIN EN 16258, a European standard for the calculation and declaration of energy consumption and GHG of transport services. The emissions balances are a function of the energy consumption by the vehicle and emission factors (g_x) that represent the emissions of the included process per unit of the final fuel. A database was created that contains the WTT and TTW factors of the different fuel and powertrain combinations. An important source of data, especially for fuel provision, is the WTW analysis that is regularly carried out under the umbrella of the JEC (JRC-Eucar-Concawe). The JEC is a collaboration between the JRC, the European Council for Automotive R&D (EUCAR) and the Oil Companies' European Organization for Environment, Health and Safety (CONCAWE). This report, which is published by the JRC, is considered as the work of reference for WTW analyses in the European

[4]Electricity from wind energy, for instance, was considered to be obtained from a practically unlimited energy source.

[5]The JEC WTW analysis consists of a WTT Report (Edwards 2014), a TTW Report (Huss 2013), and a WTW Report (Edwards 2014), which integrates the findings from the WTT and the TTW Report.

Table 3.2 Main data sources

Factor	Source
$g_{wtt(fuel)}$	(Edwards 2014) (Moro 2018)
$g_{ttw(fuel)}$	(Edwards 2014) (Hasenbichler 2018)
$g_{wtt(AdBlue)}$	(Sánchez and Antonio 2012)
$g_{ttw(AdBlue)}$	Calculated on the basis of (IPCC Guidelines for National Greenhouse Gas Inventories—Energy—Stationary Combustion 2006)
$g_{wtt(losses)}$	(Kühnel 2018) and (Wietschel 2017)

$g_{wtt(losses)}$ refers to the losses that occur within the overhead infrastructure required for supplying CEVs with electricity. In addition, the aerodynamic drag of the pantograph was accounted for as a mark-up on the vehicles' energy consumption

context and investigates a wide range of fuel and powertrain combinations. Besides the overall WTW GHG balances of the evaluated combinations, the JEC Reports also provide the disaggregated data of the included WTT pathways. In addition to the JEC Reports, other literature sources were used (see Table 3.2).

The most recent version of the JEC Report applies 2011 data to represent the EU electricity mix. To account for the continuous decarbonization trend of the EU electricity supply, data from (Moro 2018) (2013 values) was used to model the electricity-based fuel pathways (SNG_mix, H2_mix, E_mix).

3.3 Sample Application

The sample WTW analysis was conducted for a 40t semi-tractor trailer in long-haul operation, which is the vehicle type that has the highest contribution (49%) to EU CO_2 emissions within the HDV segment (Dünnebeil 2015). The assumed fuel consumption values represent a 50% payload capacity usage. The results, hereafter referred to as "WTT balances", "TTW balances" and "WTW balances" are expressed in kg CO_2e and relate to the amount of moved cargo (tkm). The GHG balances of the CI ICE using conventional diesel (80 g CO_2/tkm), hereafter referred to as diesel benchmark, were used as a baseline against which the alternative fuels and powertrains were assessed. For a better overview, the results are grouped in technologies that rely on ICEs and technologies that are based on electrified powertrains.

3.3.1 Internal Combustion Engines

The results presented in Fig. 3.4 show that all of the assessed diesel alternatives lead – with one exception—to WTW GHG savings in comparison to conventional diesel. CBG_wmr is the Biomass-to-Gas (BtG) option which achieved with 177%

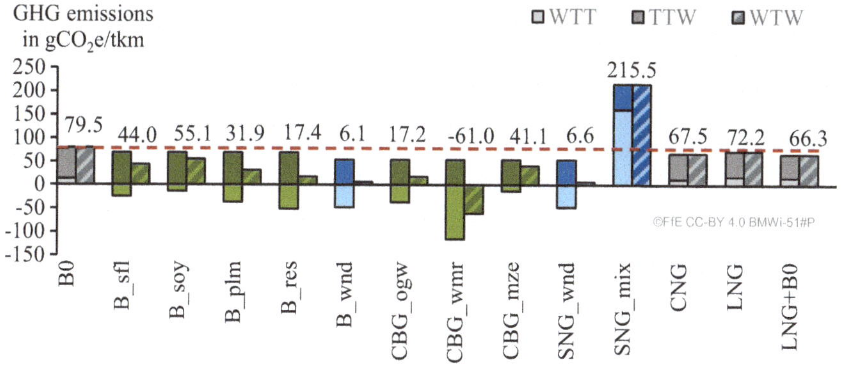

Fig. 3.4 WTW GHG emissions of ICEs

the highest GHG savings, while SNG_mix generated emissions more than two times higher than those of the diesel benchmark. Since the same energetic fuel efficiency was assumed for all the analyzed diesel options, it is mainly the variation at the WTT side that is responsible for the overall variation of the respective emissions balances. As regards PI engines, their lower thermal efficiency drives up emissions on both the WTT side and the TTW side. If used in an HPDI engine instead of a dedicated gas engine, LNG combined with a B0-diesel pilot performs better and leads to 17% WTW emission savings in comparison to the diesel benchmark.

WTT Balances

The results in Fig. 3.4 show that the investigated diesel alternatives based on RES have negative net WTT GHG balances, which is a result of the credit that was applied for the carbon absorbed during fuel production. Since electricity supply obtained from wind energy was considered carbon-free, the only GHG emissions along the WTT chain of SNG_wnd are related to fuel distribution and dispensing and are thus low enough that a negative net WTT balance was achieved after applying the carbon credit. When the same path is modeled with the EU electricity supply, however, the GHG emissions from supplying the grey electricity, which is mainly required for the energy intense electrolysis process, outweigh the carbon credit applied at the WTT side. This leads to GHG emissions that are by several orders of magnitude higher than those of the diesel benchmark.

When comparing the different biodiesel options, it is observed that their ranking in terms of their GHG emissions is largely determined by the feedstock type. Besides N_2O emissions from the usage of nitrogen fertilizers during feedstock cultivation, the productivity of the respective crops is likely to play a major role in this context. Soy, for example, which generated the highest WTT balances among the biofuel options, is characterized by a comparatively low energy productivity, meaning that biodiesel production from soy requires a relatively higher feedstock than the production from a crop with high energy productivity (Edwards 2014).

Depending on the feedstock material, the biogas yield varies significantly, which was identified as the main reason for the varying WTW balances among the different CBG options. The especially high GHG savings associated with wet manure are a result of the CH_4 emission credit that was discounted at the WTT side.[6] As CH_4 is a potential GHG, this discount was high enough to lead to a negative net WTW balance. In this context, it should be pointed out that the performance of the CBG_wmr pathway is highly dependent on the magnitude of the applied emission credit, which could be a starting-point for further discussion/analysis.

The results further illustrate that the WTT emissions from CNG and LNG provision are in the same order of magnitude than those of the diesel benchmark. CNG provision has a slightly more favorable performance than diesel, while LNG provision fares slightly less favorable. The difference between the two NG pathways is mainly traced back to the additional energy requirement of the liquefaction step. The WTT pathway for LNG provision used in the HPDI engine is the same that was used to model the dedicated gas engine. As a result of the reduced fuel consumption of the HPDI engine, WTT emission reductions of 12% (LNG + B0) in comparison to the dedicated gas engine could be achieved. The contribution of the diesel pilot, which has a more favorable WTT balance than LNG, additionally decreases the WTT balance of the HPDI pathways.

TTW Balances

As the assessed syndiesel and biodiesel alternative are used in the same engine type as the diesel benchmark, there is not a lot of potential for improvement at the TTW side. The slight difference in the TTW emissions of diesel, biodiesel, and syndiesel illustrated in Fig. 3.4 can be explained by the different fuel properties. The high carbon content and low lower heating value (LHV) of biodiesel, for instance, lead to higher emissions from fuel combustion than those of the diesel benchmark. Syndiesel, on the other hand, is characterized by a higher LHV and lower fuel carbon content and thus emits less GHG during vehicle operation. Since NG, CBG and SNG have lower carbon contents[7] than diesel (−23%, −25% and −25% based on LHV, respectively), TTW emission savings were achieved despite of the lower thermal efficiency of the dedicated gas engine in comparison to the diesel engine. In the case of the HPDI engine, the lower carbon content of NG leads to a TTW emission advantage of 8.9% for LNG + B0 in comparison to the diesel benchmark, although both propulsion systems were assumed to have identical thermal efficiencies.

[6] Methane formation naturally occurs when wet manure is left unused as landfill or applied as liquid manure (Edwards 2014).

[7] Carbon content figures were taken from (Edwards 2014), Appendix 1 & 4.

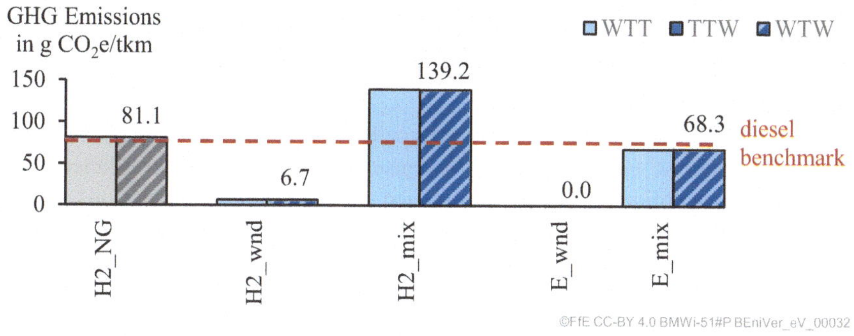

Fig. 3.5 WTW GHG emissions of electrified powertrains

3.3.2 Electrified Powertrains

Depending on the electricity that is used as power source, electrified powertrains achieve highly varying performances. Although FCEVs have the advantage that they do not generate emissions during fuel conversion and are more energy efficient than combustion engines, only one of the three analyzed hydrogen pathways achieved GHG savings in comparison to the diesel benchmark. As illustrated in Fig. 3.5, the said reductions could be achieved by electricity-based hydrogen via wind energy (H2_wnd) (92%), while hydrogen produced from the EU electricity (H2_mix) supply generated 75% higher emissions than the diesel benchmark. The deployment of CEVs, by contrast, already achieves reductions when the required electricity is supplied from the EU-mix (14%). However, the utilization of renewable energy to power CEVs is far more beneficial.

WTT Balances

The reason for the unfavorable WTT balances of the FCEV is the high energy input required for the extraction of hydrogen from the source material. Regarding electrolytic hydrogen, the same mechanisms as described for the SNG pathways apply (i.e. the high energy requirement of electrolysis drives up the emissions balances). Modeled with electricity from wind energy (H2_wnd), electrolytic hydrogen is the option with the most favorable emission performance among the investigated hydrogen pathways. The only GHG emissions associated with this path are from hydrogen compression and dispensing at the retail site, since these steps were modeled with the EU electricity supply. H2_mix, by contrast, generates emissions that are by several orders of magnitude higher than those of the diesel benchmark. The synthesis of hydrogen via steam reforming of NG is an energy intense process as well; however, H2_NG still shows lower GHG emissions than H2_mix.

Since the supply of electricity from RES was assumed to be provided as pure wind energy and to have zero GHG emissions, the direct use of electricity from RES achieves extremely beneficial WTW balances. The low efficiencies of conventional power plants, on the other hand, render the WTT performance of CEVs using E_mix by several orders of magnitude higher than those of the fossil comparator.

TTW Balances.

On the TTW side, FCEVs show an excellent performance in comparison to combustion engines. First, less energy is consumed due to the higher efficiency of the fuel cell. In addition, FCEVs were assumed not to generate any emissions during vehicle operation, meaning that WTT emissions equal WTW emissions. However, it should be pointed out that the emissions balances do not reflect any reduction of usable volume in comparison to a conventional diesel HDV. Like FCEVs, CEVs are characterized by higher efficiencies than combustion engines and zero TTW emissions. When modeled with electricity supply from the EU-mix, the beneficial TTW performance suffices to outweigh the emissions at the WTT side, leading to a WTW benefit in comparison to the diesel benchmark.

3.3.3 Sensitivity Analysis

Electricity Mix

Especially in the case of CEVs and PtX fuels, the electricity mix used to model the WTT balances plays a major role for the WTW emissions balances. The effect of a variation in the electricity mix was assessed by applying emission factors for several national electricity mixes to CEVs. The national mixes were modeled for the base year 2017 at an LCA scope (i.e. manufacture and disposal of power plants is included) with data taken from Eurostat and the ecoinvent database.

As illustrated in Fig. 3.6, GHG savings vary substantially for different Member States since the generated emissions depend on the carbon intensity of the national electricity mixes, as also pointed out by (Moro 2018). This regional variability is especially noteworthy since freight transport is unevenly distributed across the EU,

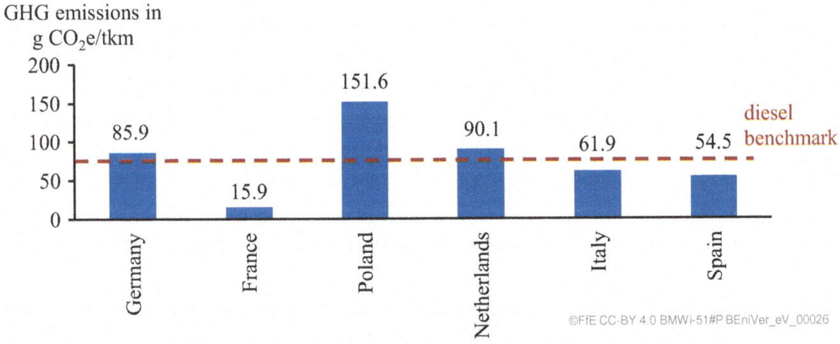

Fig. 3.6 CEV, sensitivity on variation in the electricity mix

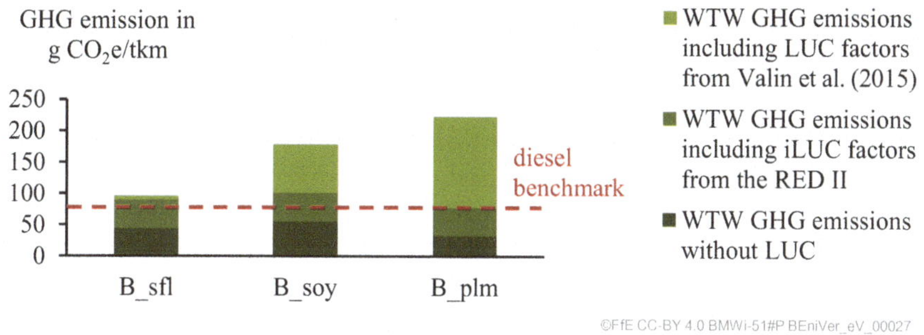

Fig. 3.7 Sensitivity analysis on LUC effects

with Poland and Germany being the main transit countries with 18% and 16%, respectively (Road Freight Transport Statistics 2018). The actual performance of a typical freight operation might thus be very different from that of an operation modeled with the average EU electricity mix. In this context, also interdependencies and the impact of an increasing electricity demand from the transport sector on the overall electricity system should be taken into consideration.

Land Use Change

As already mentioned, the GHG emission results presented this far do not account for emissions from LUC effects as their magnitude is linked to significant uncertainties. To account for possible effects of LUC, emission factors as stated in the Renewable Energy Directive (RED) II (Directive (EU) 2018) and an EC commissioned study (Valin 2015) were applied to different crop-based biofuel pathways. Figure 3.7 compares the WTW GHG balances without LUC to the balances that were obtained under consideration of the LUC emission factors taken from the two sources.

The results presented in Fig. 3.7 demonstrate that emissions from LUC offset the emission advantage of the crop-based biofuel pathways, although the LUC emission factors provided by the two sources differ greatly. When calculated with the emission factors as stated in the RED II, the impact of LUC on the WTW result is less drastic; but it should be noted that the LUC emission factors in the RED II only account for indirect LUC (iLUC) effects.[8] The RED iLUC factors represent averaged values per crop group, which leads to an emission increase by the same factor for each biodiesel pathway. The factors by the EC, on the other hand, individually account for the direct and indirect LUC effects related to each crop type. After applying the EC factors, palm fruit, which was found to be the crop-based biofuel with the most favorable WTW performance, is the biofuel option that causes the highest emissions.

[8]iLUC occurs, if crops for biofuels are grown on already existing agricultural lands, which may lead to new crop production on previously uncultivated areas. Direct LUC, on the other hand, occurs if crops are directly grown on previously uncultivated land (Marelli 2015).

3.4 Conclusion and Outlook

Taking feedstock availability of biofuels from waste streams, energy efficiency and LUC effects into account, the direct use of electricity in CEVs is identified as the option that yields the highest potential for reducing GHG emissions of heavy-duty long-haul transport. However, the performance of CEVs nevertheless depends on the extent to which the required electricity comes from non-fossil resources and varies across the EU Member States. Assuming a proceeding decarbonization of the EU electricity supply and technical improvements of fuel synthesis, the performance of PtX fuels is expected to increase in the long-term, but the overall low efficiency of PtX WTT pathways means that the direct use of electricity is preferable. Thus, PtX fuels are viable alternatives in those cases where the application of CEVs is not possible. Hydrogen used in FCEVs is the most favorable PtX option since a completely carbon free fuel cycle can be achieved, however given that NG is a widely available and cost-efficient feedstock and the required infrastructure is better established, the barriers to a widespread adoption are lower. The easiest route is therefore to use LNG in HPDI applications in the present time, transitioning to synthetic LNG (SLNG) or to hydrogen when electrolysis technology is more advanced and the electricity supply has been sufficiently decarbonized. It is recommended to additionally exploit the potential of fuels from existing waste streams, provided that they are not required as an ecosystem service.

In sum, the findings of this study suggest that a mixture of different fuel and powertrain combinations is the most suitable option for long-term decarbonization of the heavy-duty transport sector, with the direct, and to a lesser extent, indirect use of electricity being the most important factor. Given the projected increase in road freight demand and the high energy requirement of some alternative fuels, the rate of renewable power plant constructions within the EU must increase quickly. Thus, transitioning the energy sector is an important perquisite for decarbonizing the road freight sector.

Since the potential for GHG reductions cannot be generalized for specific fuels or powertrains and was shown to depend on a multitude of aspects, it is recommended to shift the focus at the legislative level from the currently adopted TTW scope to a WTW scope, or even better to an LCA scope in order to capture the impact of the entire life cycle. Otherwise, emissions may not be reduced, but merely shifted to another sector, i.e. the electricity sector. At the worst, the transition to alternative fuels and powertrains could even lead to increased emissions compared to the diesel benchmark if the focus is not shifted from the TTW to the WTW scope. Further research on this topic is carried out in the context of the accompanying research project BEniVer, which investigates the energy turnaround in transport from a system perspective (Laufendes Projekt: BEniVer - Begleitforschung Energiewende im Verkehr. https://www.ffe.de/beniver. 2019).

Acknowledgements This work was partially performed within the project BEniVer, which is funded by the Federal Ministry of Economic Affairs and Energy on the basis of a resolution of the German Bundestag under the funding reference 03EIV116C. The authors further thank MAN Truck & Bus, and especially Henryk Gräser, for the funding and support of this work.

References

Alamia A et al (2016) Well-to-wheel analysis of bio-methane via gasification. In: Heavy Duty Engines within the Transport Sector of the European Union. In: Applied Energy 170/2016. Elsevier, Amsterdam

Carbon Dioxide Emissions from Europe's Heavy-Duty Vehicles (2018) Copenhagen: European Environment Agency (EEA)

Commission Regulation (EU) 2017/2400 implementing Regulation (EC) No 595/2009 of the European Parliament and of the Council as regards the determination of the CO_2 emissions and fuel consumption of heavy-duty vehicles. Issued on December 12, 2017. Version of December 29, 2017; Brussels: The European Commission, 2017

Conrad J, Fattler S, Regett A et al (2019) Dynamis—Dynamische und Intersektorale Maßnahmenbewertung zur Kosteneffizienten Dekarbonisierung des Energiesystems. München: Forschungsstelle für Energiewirtschaft e.V (2019) Forschungsgesellschaft für Energiewirtschaft GmbH, Technische Universität München

Delgado O et al (2015) Assessment of heavy-duty natural gas vehicle emissions: implications and policy recommendations—White Paper. The International Council on Clean Transportation (ICCT), Berlin

dena-Leitstudie Integrierte Energiewende - Zwischenfazit Impulse und Erkenntnisse aus dem Studienprozess (2017) Deutsche Energie-Agentur GmbH (dena), Berlin

Directive 2014/94/EU of the European Parliament and of the Council on the Deployment of Alternative Fuels Infrastructure. Issued on October 22, 2014, Version of November 18, 2017; Brussels: European Parliament and the Council, 2017

Directive (EU) 2018/2001 of the European Parliament and of the Council of 11 December 2018 on the Promotion of the Use of Energy from Renewable Sources (RED). Issued December 11, 2018. The European Parliament and the Council, Brussels, Belgium, 11

Dreier D et al (2018) Well-to-wheel analysis of fossil energy use and greenhouse gas emissions for conventional, hybrid-electric and plug-in hybrid-electric city buses in the BRT system in Curitiba, Brazil. In: Transportation research Part D: transport and environment 58/2018. Elsevier, Amsterdam

Dünnebeil F et al (2015) Zukünftige Maßnahmen zur Kraftstoffeinsparung und Treibhausgasminderung bei Schweren Nutzfahrzeugen. Umweltbundesamt (UBA), Dessau-Roßlau

Edwards R et al (2014) Well-to-tank report version 4.a—JEC well-to-wheels analysis in: well-to-wheels analysis of future automotive fuels and powertrains in the European context. European Commission, Joint Research Centre, Institute for Energy and Transport, Luxembourg

Edwards R et al (2014) Well-to-wheels report version 4.a—JEC well-to-wheels analysis of future automotive fuels and powertrains in european context. European Commission, Joint Research Centre, Institute for Energy and Transport, Luxembourg

Harrison P et al (2014) Wasted: Europe's untapped resource—an assessment of advanced biofuels from wastes & residues. The International Council on Clean Transportation (ICCT), Berlin

Hasenbichler G (2018) HDGAS—Heavy duty GAS engines integrated into vehicles

Huss A et al (2013) Tank-to-wheels report version 4.0—JEC well-to-wheels analysis. European Commission, Joint Research Centre, Institute for Energy and Transport, Luxembourg

IPCC Guidelines for National Greenhouse Gas Inventories—Energy—Stationary Combustion (2006) Institute for Global Environmental Strategies, Kamiyamaguchi Hayama, Kanagawa

Isenstadt A et al (2017) Diesel engines—Working Paper. The International Council on Clean Transportation (ICCT), Berlin

Kirsten K et al (2016) Alternative fuels in the well-to-wheel debate. In: ATZextra worldwide 21(11)/2016. Springer Fachmedien, Wiesbaden
Koffi B et al (2017) CoM default emission factors for the Member States of the European Union. European Commission, Brussels
Kühnel S et al (2018) Oberleitungs-Lkw im Kontext Weiterer Antriebs- und Energieversorgungsoptionen für den Straßengüterfernverkehr - Ein Technologie- und Wirtschaftlichkeitsvergleich. Öko-Institut e.V., Berlin
Laufendes Projekt: BEniVer - Begleitforschung Energiewende im Verkehr. https://www.ffe.de/beniver. Retrieved 21 Jan 2019
Marelli L et al (2015) The impact of biofuels on transport and the environment, and their connection with agricultural development in Europe—Policy Department B, Structural and Cohesion Policies. European Commission, Brussels
Moro A et al (2018) Electricity carbon intensity in European Member States: impacts on GHG emissions of electric vehicles. In: Transportation Research Part D 64/2018. Elsevier, Amsterdam
Neuling U et al (2017) Conversion pathways toward transportation fuels: identification and comparison. In: Encyclopedia of sustainability science and technology, Robert A. Meyers. Springer, New York
Pachauri R et al (2007) Climate Change 2007: synthesis report. Contribution of Working Groups I, II and III to the Fourth Assessment Report of the Intergovernmental Panel on Climate Change. The Intergovernmental Panel on Climate Change (IPCC), Genf
Questions and Answers on the Commission Strategy for Reducing Heavy-Duty Vehicles' (HDVs) Fuel Consumption and CO_2 Emissions [Press Release] (2014) European Commission, Brussels
Report of the Conference of the Parties on its Nineteenth Session, Held in Warsaw from 11 to 23 November 2013, Addendum, Part Two: Action Taken by the Conference of the Parties at its Nineteenth Session (FCCC/CP/2013/10/Add.3). Bonn: United Nations Framework Convention on Climate Change (FCCC), 2014
Road Freight Transport Statistics (2018) Eurostat, Luxemburg. https://ec.europa.eu/eurostat/statistics-explained/index.php?title=File:Road_freight_transport-02.jpg
Sánchez G, Antonio J et al (2012) Comparison of life cycle energy consumption and GHG emissions of natural gas, biodiesel and diesel buses of the Madrid transportation system. In: Energy 47(1)/2012. Elsevier, Amsterdam
Schaub G, Pabst K, Lüft M, Velji A (2007) Neuartige Kraftstoffe und Zukünftige Abgasemissionen bei Kraftfahrzeugen. Universität Karlsruhe, Karlsruhe
The Future of Trucks—Implications for Energy and the Environment (2017) International Energy Agency (IEA), Paris
Too Big to Ignore—Truck CO_2 Emissions in 2030 (2015) Transport and Environment (T&E), Brussels
Valin H et al (2015) The land use change impact of biofuels consumed in the EU. European Commission, 2015
Wietschel M et al (2017) Machbarkeitsstudie zur Ermittlung der Potentiale des Hybrid-Oberleitungs-Lkw - Studie im Rahmen der Wissenschaftlichen Beratung des BMVI zur Mobilitäts- und Kraftstoffstrategie. Karlsruhe: Fraunhofer Institute for Systems and Innovation Research (Fraunhofer ISI)

Chapter 4
Pathways and Environmental Assessment for the Introduction of Renewable Hydrogen into the Aviation Sector

Christina Penke, Christoph Falter, and Valentin Batteiger

Abstract The availability of low-cost renewable electricity from solar and wind energy provides an opportunity to use renewable hydrogen for aviation fuel production. The Power-to-Liquid (PtL) pathway provides a perspective to synthesize large amounts of sustainable jet fuels from H_2O and CO_2. The environmental performance of biofuel pathways can immediately be improved by the use of hydrogen from renewable energy. In the long-term perspective, it is conceivable that hydrogen can also be used directly as fuel in future hydrogen aircraft. We present an approach that allows comparative statements about the greenhouse gas balance of future hydrogen supply options for aviation. The model for direct hydrogen use takes into account the production of liquid hydrogen and its logistics. The reference cases are fuel production using renewable hydrogen in biorefineries and fuel production by PtL. We discuss future scenarios for an intensive use of hydrogen from renewable energy in aviation and evaluate the resulting environmental impacts.

Keywords Prospective life cycle assessment · Renewable fuels · Hydrogen · Aviation

4.1 Introduction

During the last decade, electricity generation from solar radiation and wind power became the most competitive electricity generation technologies in many regions of the world (IRENA 2018). This enables the further decarbonization of the energy and transport sector. Such schemes are frequently termed "sectorial coupling" or "Power-to-X" (PtX). Hydrogen from water electrolysis is usually an intermediate product for PtX pathways towards transportation fuels and renewable chemicals. This article assesses selected options to utilize renewable hydrogen, i.e. hydrogen produced from renewable energy, for the production of renewable aviation fuels.

C. Penke (✉) · C. Falter · V. Batteiger
Bauhaus Luftfahrt e.V., Willy-Messerschmitt-Str. 1, 82024 Taufkirchen, Germany
e-mail: Christina.penke@bauhaus-luftfahrt.net

S. Albrecht et al. (eds.), *Progress in Life Cycle Assessment 2019*,
Sustainable Production, Life Cycle Engineering and Management,
https://doi.org/10.1007/978-3-030-50519-6_4

The so-called Power-to-Liquid pathway (PtL), which synthesizes "drop-in" capable jet fuel from renewable H_2 and CO_2, provides a credible option for a complete substitution of conventional jet fuel with minor adaptions within the aviation sector (Schmidt et al. 2018). Within this work, the PtL process serves as a benchmark to assess further PtX schemes for aviation from a prospective life-cycle perspective. Namely, the improvement of the sustainability of biofuels by a substitution of the current H_2 source within their production and the option to use H_2 directly as a fuel for aircraft. Both options are assessed for two exemplary cases: the substitution of conventional H_2 in the production of HEFA fuel (hydrotreated esters and fatty acids), and the use of liquid hydrogen (LH_2) at cryogenic temperatures for future long-haul aircraft. It is important to note that the comparability of the investigated options is limited. The production of refinery hydrogen via water electrolysis can be deployed immediately, while liquid hydrogen powered aircraft require substantial development effort. Rather than a competitive comparison, we intend to prepare a comprehensive assessment of various PtX schemes by spanning the full range of possibilities from a gradual replacement of refinery hydrogen to potential liquid hydrogen concepts in the long-term future.

4.2 Reviewing the Options for the Use of Hydrogen from Renewable Electricity in Aviation

4.2.1 Power-to-Liquid

A schematic overview of selected options to produce aviation fuels with renewable electricity input is given in Fig. 4.1. The Power-to-Liquid process (Schmidt et al. 2018; König et al. 2015), which combines H_2 from water electrolysis with CO_2 as a source of carbon serves as a benchmark for the production of jet fuel from renewable electricity. A fraction of the H_2 stream is consumed during the conversion of CO_2 to CO via the reversed water gas shift reaction. The resulting synthesis gas stream with a molar H_2/CO ratio of about 2 is converted to liquid hydrocarbons via a Fischer-Tropsch process. Schmidt et al. (2018) argue that one key requirement for sustainable PtL production is low-carbon electricity (GHG intensity well below 100 gCO_2eq/kWh_{el}). Another essential requirement is a sustainable CO_2 source (Falter et al. 2016). Sustainable CO_2 point sources can be utilized to limited extent, but scaling synthetic fuel production towards future aviation fuel demand requires scalable CO_2 sources, such as direct air capture, to provide several Gt_{CO2} per year (Batteiger et al. 2018). The PtL process can potentially produce very large amounts of synthetic jet fuel from H_2O and atmospheric CO_2, therefore it can serve as a meaningful benchmark to select and discuss less elaborate PtX schemes, such as the production of refinery H_2, as well as future, more efficient PtX schemes, such as the direct use of H_2 as an aviation fuel.

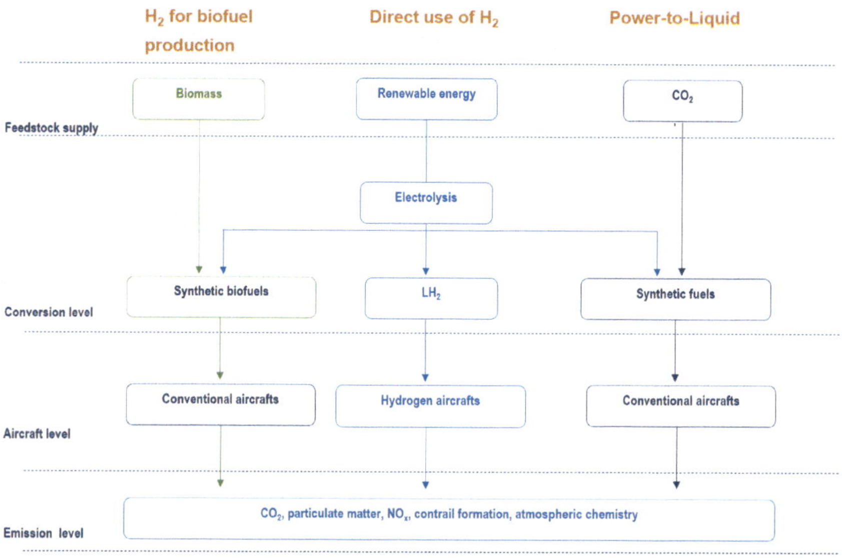

Fig. 4.1 Overview of basic options to use renewable electrolysis hydrogen for aviation fuel production

4.2.2 *Hydrogen for Biofuel Production*

A great fraction of the current H_2 production is consumed within refineries, for processes such as hydrotreating, hydrogenation, hydroisomerization, hydrocracking or desulphurization (de Klerk 2011). Within petroleum refineries, H_2 is frequently produced by reforming of hydrocarbons. Obviously, renewable H_2 can substitute conventional H_2 sources and thereby reduce the GHG footprint of refinery products. Similarly to petroleum refining, aviation biofuel pathways involve hydrogen provision to produce drop-in capable jet fuel. This is different to the most common biofuels for road transportation (ethanol and biodiesel), which do not necessarily consume H_2 for their production.

Hydroprocessed esters and fatty acids (HEFA) or hydrotreated vegetable oils (HVO) are currently the most common type of aviation biofuels, which is why a HEFA pathway was selected for further analyses in Sect. 4. HEFA fuels are produced from plant oils or other lipids and directly involve H_2 for production. HEFA fuels are already approved as a blend component to conventional jet fuel at a maximum blending ratio of 50/50 (ASTM D7566-19 2019). Further pathways towards renewable jet fuel include pyrolysis, hydrothermal liquefaction (HTL), alcohol-to-jet (AtJ), and direct sugar to hydrocarbons (DSHC), and involve different specific amounts of hydrogen for their production (De Jong et al. 2017).

Most biofuel pathways consume H_2 mainly during the final upgrading steps. This is different for Biomass-to-Liquid (BtL) pathways, where biomass is gasified to produce a syngas for Fischer-Tropsch conversion. The syngas composition of biomass gasification depends on the feedstock composition and the chosen gasification technology (Rauch et al. 2014). Steam gasification yields a syngas with relatively high H_2/CO ratio. Oxygen-based gasification schemes are more energy efficient, but the H_2 content of the syngas is lower and an O_2 source is required. Renewable H_2 can adjust the composition of the H_2 deficient syngas towards the desired ratio and O_2 evolving from water electrolysis can potentially serve as an input for biomass gasification. One particularly appealing case is the combination of entrained flow gasifiers, which yield a low H_2/CO ratio, but offer technological benefits (Rauch et al. 2014).

4.2.3 Direct Use of Hydrogen as Aviation Fuel

The specific energy density of hydrogen (LHV: 33.3 kWh/kg, HHV: 39.4 kWh/kg) is about three times as large as that of conventional jet fuel, but the volumetric energy density is poor. Liquid hydrogen (LH_2) has a density of only 70.9 g/L, which equates to an energy density (LHV: 2.36 kWh/L, HHV: 2.8 kWh/L) 75% lower than hydrocarbon fuels. Compressed H_2 takes another factor of two (700 bar technology) to four (350 bar technology) as much volume as LH_2 and the gravimetric capacity of pressure tanks is in the range of 5–6%$_{\text{weight}}$ (Hua et al. 2011). Compressed H_2 has been used in small aircraft (Rathke et al. 2013), but it is commonly accepted that the use of LH_2 is beneficial for larger hydrogen aircraft, especially for long-haul concepts.

The EU project Cryoplane (2000 to 2003) Cryoplane (2003) investigated different LH_2-fuelled transport aircraft concepts. The concept study found an increased energy consumption per passenger of 9% (long range) to 34% (business jet) compared to conventional reference aircraft. In contrast, a study by the Hamburg University of Applied Sciences found that a LH_2 fuelled freighter aircraft could be 7% more efficient than a conventional reference (Seeckt et al. 2010). A recent concept study by Bauhaus Luftfahrt (2019) found an energy consumption for a long-haul transport aircraft that is at par with a conventional reference aircraft (Troeltsch 2019), where both the LH_2 aircraft and the conventional reference aircraft were designed for an entry into service in 2035 and a reduced cruise speed of Mach 0.7. The individual studies evaluate the hydrogen fuel demand of different aircraft categories at different technological state of art and with different sets of assumptions. The differences in the findings from the respective studies remain within a reasonable variance of results. Therefore, it can be concluded that the energy demand at aircraft level is similar to the conventional reference for a broad range of hydrogen fueled aircraft categories.

A common assumption of the above-mentioned studies is that LH_2 is combusted in an adapted aircraft turbine. The utilization of fuel cells may provide additional efficiency and emission benefits; fuel cell propulsion is demonstrated by DLR at small aircraft level (H2FLY Homepage (Online) 2019).

4.2.4 Climate Impact of Non-CO_2 Emissions at Cruise Altitude

An important consideration, both for the direct use of hydrogen as an aviation fuel, and also for the use of various synthetic jet fuels is the climate impact of "non-CO_2 effects" at cruise altitude. The net GHG emissions resulting from fuel production and combustion act as well-mixed greenhouse gases, i.e. the climate impact does not significantly depend on the location of the emission. Consequently, carbon uptake during fuel production can be balanced against CO_2 release during jet fuel combustion. In contrast, the climate impact of NO_x, SO_x, soot and H_2O in form of water vapor, depend significantly on the location and the ambient conditions at cruise altitudes. A detailed review of the full climate impact of aviation is e.g. given by Grewe et al. (Grewe et al. 2017), while within the scope of this manuscript we briefly discuss important considerations regarding the differences in the non-CO_2 effects due to alternative fuels use.

A quantitative comparison of the radiative forcing or net warming resulting from non-CO_2 effects is complex. Several studies find an additional warming effect that is comparable or larger in magnitude compared to the CO_2 effect resulting from conventional jet fuel combustion (H2FLY Homepage (Online) 2019; Kärcher 2018). Several renewable fuel pathways, including the PtL benchmark, are capable of reducing net GHG emissions by more than 80% (see Fig. 4.3). For climate change mitigation it is important to reduce the non-CO_2 effects caused by sustainable aviation fuels compared to conventional jet fuel, in parallel to the decarbonization of aviation fuel.

The use of renewable jet fuel can reduce the non-CO_2 effects of aviation (Burkhardt and Bock 2018) mainly due to a reduction of soot and aerosol emissions that affect the properties of contrails and contrail cirrus, while the formation of NO_x is not significantly affected by the composition of drop-in fuels (Schripp et al. 2018). The aromatics content is a key parameter to reduce the non-CO_2 effect of renewable jet fuels (Kärcher 2018). The combustion of H_2 is expected to reduce aerosol emissions below ambient levels and can also significantly reduce NO_x formation, however, the specific emission of water vapor is increased compared to the combustion of hydrocarbon fuels. Ponater et al. have assessed the potential to reduce the climate impact of aviation with hydrogen aircraft, and found significant reductions in the radiative forcing due to reduced NO_x emissions and the formation of fewer but larger ice crystals in cryoplane contrails due to the absence of particles in the engine exhaust (Ponater et al. 2007).

Despite of the importance for the full climate impact of aviation, we are not aware of an established methodology to include non-CO_2 effects in a life-cycle assessment of renewable jet fuel production. More research is needed to quantify the impact of alternative fuel use, especially in case of non-drop in fuel use such as hydrogen.

4.3 Methodical Approach

In order to estimate the environmental impact resulting from using renewable hydrogen in aviation, the methodology applied for this study is described in the following.

4.3.1 Goal and Scope

The aim of this study is to assess the greenhouse gas emission reduction that can be achieved by using renewable hydrogen from water electrolysis for fuel production. For this purpose, renewable fuel production pathways with different technological maturity are considered, in which hydrogen is used for processing and refining. The potential to reduce greenhouse gases is evaluated during the production and use phases by mapping material and energy flows for the different fuel paths. In a subsequent discussion, the technological maturity of the respective category is taken into account by considering the reduction potential by using renewable electrolysis hydrogen in relation to the time of future development.

4.3.2 Impact Categories

The environmental impact category is considered to be the greenhouse gas reduction potential resulting from fuel production and the use phase equivalent to combustion aircraft turbines. It is assumed that the combustion of conventional jet fuel and its synthetic counterpart are of equal detrimental effect to the environment and therefore do not represent a basis for differentiation.

To evaluate the climate impact this study focuses on the GHG emissions. The GHG emissions considered were CO_2, CH_4 and N_2O applying their 100-year global warming potential.

Models (Cryoplane 2003; Troeltsch 2019) have shown that NO_x emissions of hydrogen aircraft can be lower than those of a conventional aircraft. Experiments in which hydrogen is burned in stationary gas turbines have shown NO_x emissions below 20 ppm per converted amount of fuel for different payload ranges (Tekin 2015).

The use of hydrogen is also expected to result in fewer aerosols and soot particles, thus having a less impact on the climate. Since the effect of aerosols on the climate has not been clearly determined, the formation of soot and aerosols is not considered. In this study, we focus quantitatively only on GHG emissions and consider the environmental impacts in terms of further impact categories in the discussion.

4.3.3 Life-Cycle Inventory

In order to compare drop-in capable fuels with hydrogen, which requires the development of new turbine and aircraft concepts, we select the metric MJ/(passenger km) as functional unit for this study. GHG analysis for several production pathways for jet fuel was performed using GREET (Version 2017). Unless otherwise stated, the process paths defined in the GREET model was used. In addition, Ecoinvent (v3.5) was applied to describe the losses in hydrogen transport in terms of environmental impact. The system boundaries considered for this study are shown in Fig. 4.2.

The Well-To-Pump system boundaries include the provision of the feedstock, the proper fuel production process and the transport and storage of the fuels. The Well-To-Wake boundaries also take the use phase into account by the combustion of hydrocarbons or hydrogen in aircraft turbines.

For the evaluation of the by-products, an allocation of GHG emissions based on the energy content was chosen, since in addition to the main product jet fuel, mainly other hydrocarbon fractions are produced, that are primarily used for other energy purposes.

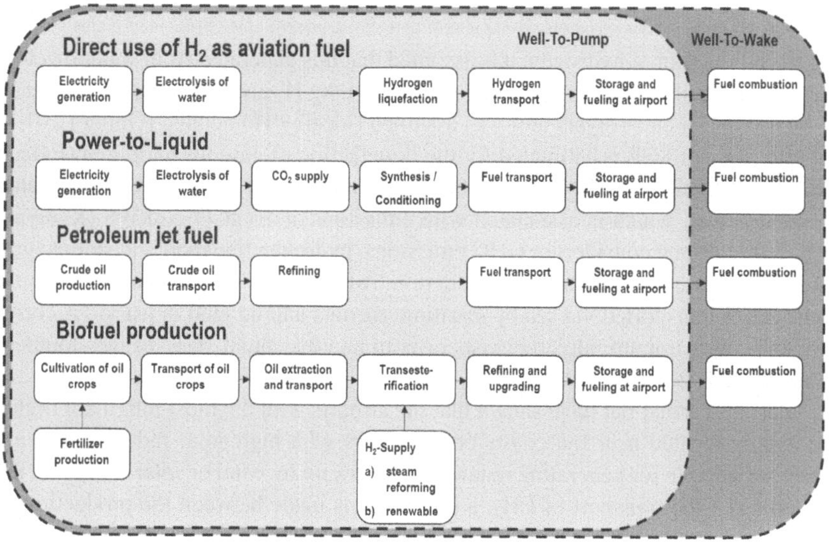

Fig. 4.2 System boundaries for the different considered fuel pathways: direct use of hydrogen as aviation fuel, PtL, petroleum jet fuel and biofuel production

4.4 System Modelling

In order to compare the different fuel pathways of petroleum fuel, PtL, biofuel and direct use of hydrogen as aviation fuel, these were each modelled within the system boundaries Well-To-Wake. To represent the reference, the "Conventional Jet Fuel from Crude Oil" model available in GREET was used.

4.4.1 Biofuel Production

Soy oil was selected as feedstock for biofuel production via HEFA. The processing of crop oil to jet fuel involves hydrotreatment to deoxygenate the oil with subsequent hydrocracking. As HEFA fuels are already available today, the data provided in GREET are used for this production path.

4.4.2 Direct Use of Hydrogen as Aviation Fuel

For the production of hydrogen, it is assumed that it is generated from water by electrolysis. A specific energy requirement of 48 kWh/kg H_2 and a conversion efficiency of 68.8% for an electrolysis unit was assumed (idealhy Homepage (Online) 2019). An efficiency of 86% is estimated for the liquefaction of gaseous hydrogen (E4tech Homepage (Online) 2019). It has been assumed that the electricity is generated by wind power, which is associated with emissions of 10 gCO_2eq/kWh (Kadiyala et al. 2017). When considering GHG emissions, hydrogen transport and storage was taken into account in addition to production. Additionally, it is assumed that liquid hydrogen is transported via sea by maritime carriers and on land in trucks. A corresponding infrastructure already exists or is in an conceptual stage of development (Kamiya et al. 2015).

Analyses carried out have shown that the airports with the most long-haul flights are mainly located near the coasts or in regions with high solar radiation and are therefore suitable for generating renewable electricity by wind or solar energy. In the modelling of the transport of LH_2, a difference is made between the production of hydrogen directly at the respective airports and the transport over longer distances. For this reason, two scenarios were considered as base cases. A realistic assessment lies between the two scenarios.

- Scenario A: LH_2 is produced directly on site at the airport under favorable conditions using renewable energy (wind or photovoltaics).
- Scenario B: LH_2 is produced off-site under favorable conditions, then transported to the airport over a distance of 9000 km by sea and 1000 km by truck.

For scenario B, the assumption is made that a loss of 0.2% per day is to be expected during storage. In addition, it is assumed that 0.49% losses will result per conversion stage.

4.4.3 Power-to-Liquid

For PtL process modelling, the same assumptions are made for the electrolyzer and the present renewable energy generation as for the production of renewable hydrogen. Direct air capture (DAC) with an energy demand of 1700 kWh/t CO_2 (heat and electricity) is regarded as the process for capturing CO_2. In addition, an overall efficiency of 70% is assumed for Fischer-Tropsch synthesis (König 2016).

4.4.4 Hydrogen Demand for Different Fuel Options

The fuel production pathways considered in this study have in common that hydrogen is additionally necessary for the provision of drop-in capable fuel, the specific hydrogen demand differs as shown in Table 4.1. The specific hydrogen demand is calculated in relation to the main product jet fuel and additionally in relation to all products (allocation by energy).

4.5 Results and Discussion

For scenario A, the energy demand is 54.8 kWh/kg LH_2, which corresponds to a total efficiency of 59.2% for the production of LH_2. In scenario B, taking transport

Table 4.1 Specific hydrogen demand for different aviation fuel production pathways (De Jong et al. 2017; König 2016)

Fuel pathway	Specific H_2 demand [$MJ_{H2}/MJ_{jet\ fuel}$]	Specific H_2 demand [$MJ_{H2}/MJ_{all\ products}$]	Typical feedstock
HVO/HEFA	0.15	0.12	Oils and fats
Jet fuel from pyrolysis	5.44	0.38	Flexible
Jet fuel from HTL oil	1.31	0.13	Flexible
Alcohol-to-Jet (AtJ)	0.08	0.06	Sugar and starch
Direct sugar to hydrocarbon (DSHC)	0.52	0.29	Sugar and starch
Power-to-Liquid (PtL)	3.50	1.51	CO_2 and H_2O

into account, the energy requirement is 56.3 kWh/kg LH_2. The transport of LH_2 thus accounts for a maximum of 2.5% in the total energy balance.

4.5.1 GHG Emissions for Different Fuel Production Pathways

The resulting GHG emissions are shown in Fig. 4.3. In the WTW system boundaries, compressed hydrogen with 3.8 gCO_2eq/MJ is the best of the compared fuel production pathways. However, hydrogen in this form is only conditionally suitable for future use in aviation. LH_2, which appears to be better suited has only slightly higher GHG emissions (4.6 gCO_2eq/MJ) due to the additional liquefaction stage (Scenario A). For scenario B, which also takes the transport into account, emissions of 4.7 gCO_2eq/MJ result.

The modelling of the PtL process results in GHG emissions of 19.3 gCO_2eq/MJ. For the biofuel path considered using soy oil, GHG emissions of 59.50 gCO_2eq/MJ result. These relatively high values come mainly from the soy oil production including fertilization, cultivation of agricultural land and extraction of oil.

If the fuel required for upgrading from soy oil to jet fuel is replaced by renewable hydrogen, an almost 4% reduction in GHG emissions can be achieved (57.3 gCO_2eq/MJ).

All options considered except H_2 (non-renewable) perform better than conventional fuel, which is subject to a value of 85.50 gCO_2eq/MJ.

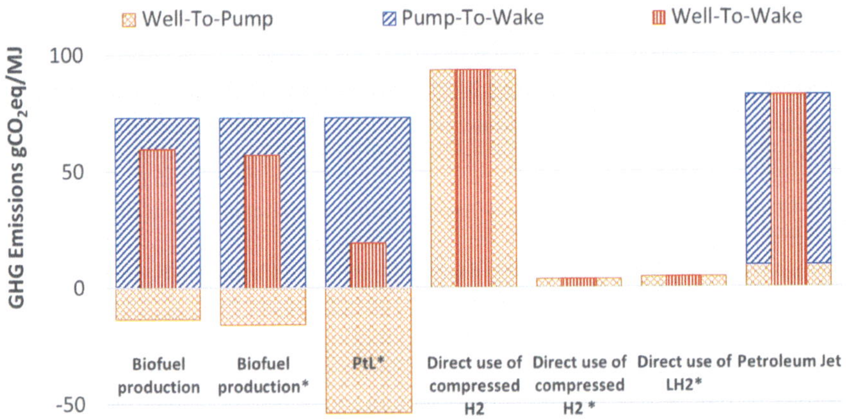

Fig. 4.3 GHG emissions of the considered jet fuel production pathways (*Hydrogen produced by electrolysis using renewable energy)

4.5.2 Discussion

When interpreting the results, it has to be taken into account that we have looked in this study at production pathways that encompass a different technical maturity. Although hydrogen from water electrolysis produced by renewable energy has the least GHG emissions, this does not mean that it is currently the best fuel option. In the short term, cost-efficiency and a low implementation threshold of HEFA biofuels gives it an advantage over other options to reduce GHG emissions. However, due to its limited sustainable production potential and limited GHG savings, its long-term overall potential for GHG reduction of aviation will be limited, even with renewable hydrogen. In long term, the costs of renewable electricity from wind and photovoltaics are expected to decrease. This would probably lead to an expansion of the LH_2 infrastructure and lower costs for LH_2 produced via water electrolysis are to be expected.

The study has also shown that hydrogen production using the current electricity mix leads to even higher GHG emissions than conventional fuel. At present, renewable energies and LH_2 infrastructure are not available as necessary for a supply of hydrogen for aviation to significantly or completely substitute for conventional jet fuel. In addition, the aircraft suitable for the use of hydrogen in the aviation sector are not yet developed. However, the progressive use of hydrogen in other areas can further advance the expansion of the infrastructure and the use of renewable energies.

Since assumptions regarding future infrastructure had to be made for modelling future technologies (PtL and production of renewable hydrogen), the results are subject to change with a variation of those assumptions. The robustness of the results with respect to the uncertain future will be evaluated in future work with sensitivity studies.

In order to take into account the different (sustainable) production potentials of the pathways considered, further studies will include the effect of limited availability of raw materials, ant its correlation with GHG emission reduction potentials. To this end, LUC or iLUC of different feedstock have to be considered.

It should also be mentioned that non-CO_2 aviation effects were not quantified in the assessment of environmental impacts. However, in order to obtain a comprehensive picture of climate impact of different fuels, it is important to quantitatively include them in analyses. Extensive models are required that describe the complex interactions of different substances in the species in order to be able to map these non-CO_2 effects in a suitable way. This, and an even better understanding of the technology options and driving factors, will provide an effective strategy for decarbonizing global aviation.

References

ASTM D7566—19 (2019) ASTM International, Standard specification for aviation turbine fuel containing synthesized hydrocarbons. www.astm.org/Standards/D7566.htm

Batteiger V, Falter C, Gálvez Martos JL, Dufour J, Iribarren D (2018) Public report: techno-economic and environmental analysis of CO_2 provision from various sources

Burkhardt U, Bock L (2018) Mitigating the contrail cirrus climate impact by reducing aircraft soot number emissions. NPJ Clim Atmos

Cryoplane—Liquid Hydrogen Fuelled Aircraft—System Analysis, GRD1-1999-10014, 2000–2003

De Jong S et al (2017) Life-cycle analysis of greenhouse gas emissions from renewable jet fuel production. Biotechnol Biofuels, 1–18

de Klerk A (2011) Fischer–Tropsch fuels refinery design. Energy Environ Sci 4(1177):1177–1205

E4tech Homepage (Online). Available: http://www.e4tech.com/. Accessed 29 Oct 2019

Falter C, Batteiger V, Sizmann A (2016) Climate impact and economic feasibility of solar thermochemical jet fuel production. Environ Sci Technol 50:470–477

Grewe V et al (2017) Mitigating the climate impact from aviation: achievements and results of the DLR WeCare project 4:34

H2FLY Homepage (Online). Available http://h2fly.de/. Accessed 29 Oct 2019

Hua TQ et al (2011) Technical assessment of compressed hydrogen storage tank systems for automotive applications. Int J Hydrogen Energy 36:3037–3049

idealhy Homepage (Online). Available: https://www.idealhy.eu/. Accessed 29 Oct 2019

IRENA (2018) Renewable power generation costs in 2018

Kadiyala A, Kommalapati R, Huque Z (2017) Characterization of the life cycle greenhouse gas emissions from wind electricity generation systems. Int J Energy Environ Eng 8(1):55–64

Kamiya S, Nishimura M, Harada E (2015) Study on introduction of CO_2 free energy to Japan with liquid hydrogen. Phys Proc 67:11–19

Kärcher B (2018) Formation and radiative forcing of contrail cirrus. Nat Commun, 1–17

König DH (2016) Technoökonomische Prozessbewertung der Herstellung synthetischen Flugturbinentreibstoffes aus CO_2 und H_2. Universität Stuttgart

König DH, Freiberg M, Dietrich R-U, Wörner A (2015) Techno-economic study of the storage of fluctuating renewable energy in liquid hydrocarbons. Fuel 159:289–297

Ponater M, Pechtl S, Sausen R, Schumann U, Hu G (2007) Potential of the cryoplane technology to reduce aircraft climate impact: a state-of-the-art assessment

Rathke P, Kallo J, Schirmer J, Stephan T, Waiblinger W, Weiss-Ungethüm J (2013) Antares DLR-H2—flying test bed for development of aircraft fuel cell systems. ECS Trans 51:229–241

Rauch R, Hrbek J, Hofbauer H (2014) Biomass gasification for synthesis gas production and applications of the syngas. WIREs Energy Environ 3:343–362

Schmidt P, Batteiger V, Roth A, Weindorf W (2018) Power-to-liquids as renewable fuel option for aviation : review

Schripp T et al (2018) Impact of alternative jet fuels on engine exhaust composition during the 2015 ECLIF ground-based measurements campaign. Environ Sci Technol 52(8):4969–4978

Seeckt K, Heinze W, Scholz D (2010) Hydrogen powered freighter aircraft—the final results of the green freighter project. In: 27th Congr Int Counc Aeronaut Sci 2010, ICAS 2010

Tekin N (2015) (KAWASAKI Gas Turbine EuropeGmbH). Entwicklung innovativer Wasserstoff-Verbrennungssysteme für Industriegasturbinen

Troeltsch F (2019) Concept for a hydrogen-powered long-haul aircraft. Bauhaus Luftfahrt symposium, 8 May 2019

Chapter 5
Life Cycle Assessment of a Polymer Electrolyte Membrane Water Electrolysis

Elke Schropp, Gabriel Naumann, and Matthias Gaderer

Abstract In this study, the environmental burdens of polymer electrolyte membrane water electrolysis (PEMWE) producing hydrogen are analyzed with the help of a life cycle assessment (LCA). Three scenarios, which differ in the techniques applied for generating electricity for hydrogen production and the operation mode, are considered. Using electricity generated with the currently available generation structure leads to the GWP100 of 32.52 kg CO_2-eq per kilogram H_2, which leaves the plant with a purity of 99.9% at 35 bar and 333–335 K. These emissions are to a large extent caused throughout the plant's operation but not during its production phase. Thus, currently a large proportion of the GWP100 results from electricity production. As electricity generation based on renewable energy continues to develop, the greenhouse gas emissions per kilowatt-hour electricity decrease. Consequently, the overall greenhouse gas emissions as well as those caused within the operation phase of PEMWE decline and the relative impact of the production phase grows. In case of operating the electrolysis plant exclusively with electricity originating from wind, the GWP100 of 0.93 kg CO_2-eq/kg H_2 results, of which more than one quarter is assigned to the production of the plant. Thus, the production phase of a PEMWE plant must be thoroughly considered when the share of renewable energy in electricity generation increases. Apart from the GWP100, other impact categories are analyzed. Comparing the three scenarios reveals that the application of renewable energy for electricity generation is beneficial for most impact categories. Nevertheless, ADP elements, FATEP inf., HTP inf. and ODP are negatively affected.

Keywords Life cycle assessment · Polymer electrolyte membrane water electrolysis · Electrolysis · Hydrogen production · Power-to-gas · Synthetic fuels

E. Schropp (✉) · G. Naumann · M. Gaderer
Technical University of Munich, Campus Straubing for Biotechnology and Sustainability, Professorship of Regenerative Energy Systems, Schulgasse 16, Straubing, Germany
e-mail: elke.schropp@tum.de

G. Naumann
e-mail: gabriel.naumann@tum.de

M. Gaderer
e-mail: gaderer@tum.de

S. Albrecht et al. (eds.), *Progress in Life Cycle Assessment 2019*, Sustainable Production, Life Cycle Engineering and Management,
https://doi.org/10.1007/978-3-030-50519-6_5

5.1 Introduction

Climate change is one of the major challenges for humankind in the 21th century. In a recently published study, the IPCC pointed out drastic consequences for the ocean and cryosphere resulting from climate change, including an unabated warming of the global ocean and the mass loss of ice sheets and glaciers (IPCC 2019). Since the ocean and cryosphere play important roles within the Earth's ecosystems, such as the absorption and redistribution of carbon dioxide and heat as well as providing food, all humans are directly or indirectly affected by changes of the ocean and cryosphere and, thus, also by climate change (IPCC 2019). This is just one of the many reasons why limiting global warming represents an important issue, which was addressed at the United Nations Climate Change Conference in Paris in 2015. At this conference, an agreement was adopted that the global warming should be kept well below 2 °C compared to pre-industrial levels (United Nations Framework Convention on Climate Change 2015).

Reaching this target requires extensive developments, one of which is the transition of the energy sector towards renewable energy (Guy et al. 2019). With respect to electricity generation, a multitude of techniques capable of generating electricity using renewable energy sources, like wind and sunlight, have been available for years. However, most of these techniques are not able to provide electricity on a continuous basis, necessitating an expansion of storage technologies (Zerrahn et al. 2018). One such suitable technology is the conversion of electricity to hydrogen with the help of a water electrolysis. This kind of storage is often referred to as power-to-gas. However, the power-to-gas process may not only consist of hydrogen production, but may cover a further processing step. In this processing step, for example, synthetic natural gas (SNG) is produced (Varone and Ferrari 2015). SNG is advantageous since a natural gas infrastructure is already available and can be used with existing technologies in the mobility and heat sector. Thus, the power-to-gas technology enables a coupling of the mobility, electricity, and heating sectors (Fuel Cells and Hydrogen Joint Undertaking 2015).

Independent from the application of hydrogen, whether directly used or further processed, it plays an important role in the transition of the energy sector towards renewable energy as well as in the decarbonization of other sectors (Staffell et al. 2019). As this transition aims at keeping climate change within certain limits, it must be determined whether the production of hydrogen with water electrolysis remains practical. For this reason, the environmental burdens associated with the production of hydrogen using a water electrolysis are analyzed with the help of a life cycle assessment. In this context, this study focuses on the electrolysis technology based on a polymer electrolyte membrane.

5.2 Polymer Electrolyte Membrane Water Electrolysis

With respect to water electrolysis, there are three main technologies: alkaline water electrolysis, high temperature water electrolysis, and polymer electrolyte membrane water electrolysis (PEMWE). While the first is a mature technology and the second is in its early stages of development, the third is gaining wide interest and being researched intensively. The overall reaction of water electrolysis is given by Eq. 1. This reaction is endothermic and must be supplied with energy (Stolten and Krieg 2011).

The simplified design of a PEMWE cell and stack is illustrated in Fig. 5.1. As can be seen, a polymer electrolyte membrane, typically consisting of Nafion, separates the two cell halves. The electrodes are directly placed on both sides of the membrane. Conventionally, iridium dioxide (IrO_2) is used as an electrocatalyst for the anode side, where water is oxidized (Eq. 2). In contrast, the hydrogen evolution reaction at the cathode side is supported by a platinum (Pt) catalyst (Eq. 3) (Bessarabov and Millet 2018).

$$H_2O \rightarrow 0.5O_2 + H_2 \tag{1}$$

$$H_2O \rightarrow 0.5O_2 + 2H^+ + 2e^- \tag{2}$$

$$2H^+ + 2e^- \rightarrow H_2 \tag{3}$$

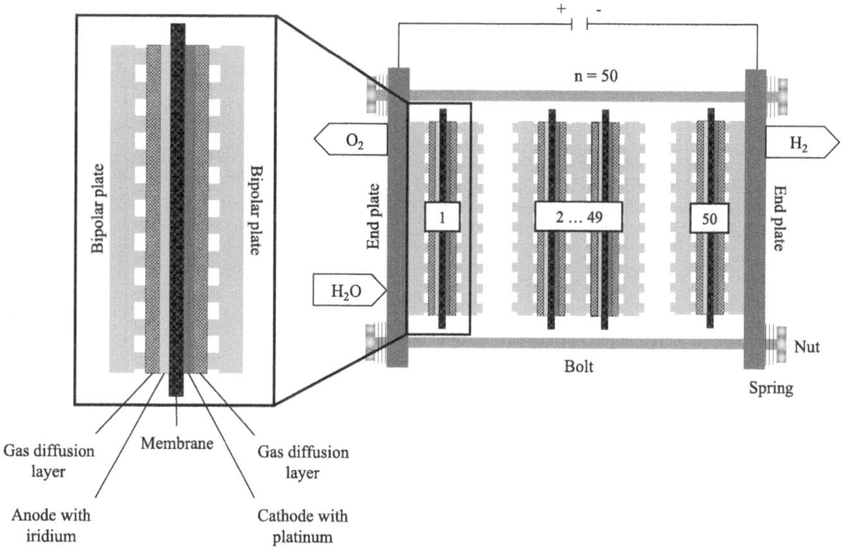

Fig. 5.1 Structure of a PEMWE cell and stack (adapted from Bareiß et al. 2019)

The combination of electrodes and membrane is called membrane electrode assembly (MEA). The MEA is sandwiched between the gas diffusion layers and the bipolar plates (Bessarabov and Millet 2018). These layers, which are also referred to as current collectors, are made of sintered titanium felt, sintered titanium powder, or expanded titanium screen mash and, thus, have a highly porous structure (Ito 2015). In an electrolysis stack, several single cells are connected in series, which results in a higher hydrogen production rate. Adjacent cells are separated by titanium bipolar plates (Wang 2015). Due to their integrated flow channels, the bipolar plates enable the transportation of both water to the cells and gases from the cells (Smolinka et al. 2015). At both ends, thick titanium end plates complete the stack. Finally, the stack is tensioned by inserting bolts between the two end plates and fixing them with springs and nuts (Millet 2015).

Figure 5.2 shows the basic configuration of the considered electrolysis plant. The main component of the plant is the electrolyzer, but further components are required for the operation of the stack, and these are known as balance of plant (BOP) components. Since the electrolyzer requires very clean water, it is treated in a purification step before inserting it into the stack, where water splitting takes place. Subsequent

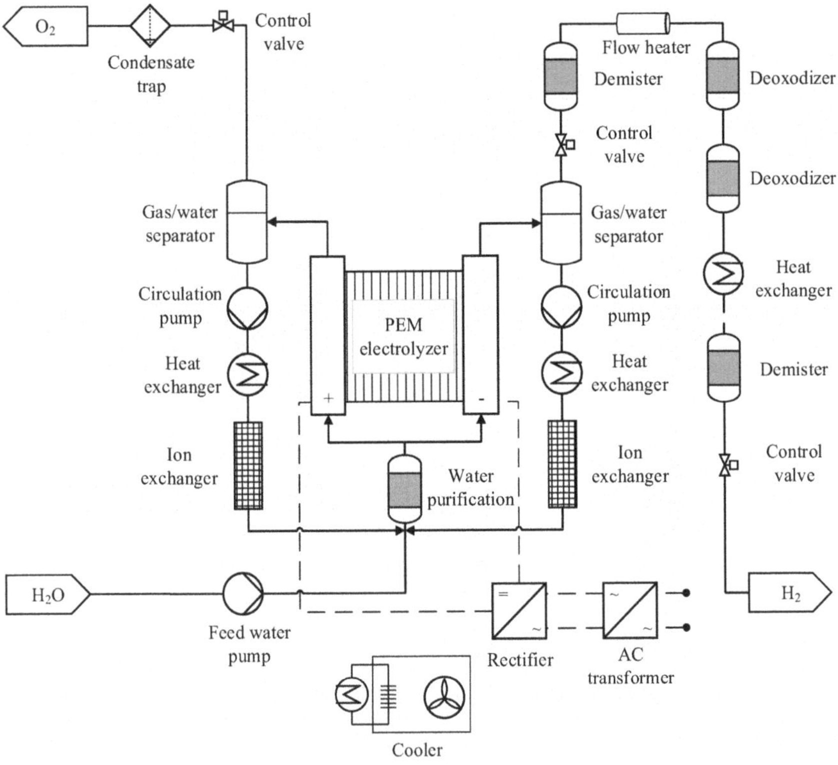

Fig. 5.2 General layout of a PEM electrolysis plant (adapted from Smolinka et al. 2015)

to the electrolyzer, gas/water separators ensure that surplus water is isolated from any produced gases. This means that some water is not split, instead passing through and cooling the electrolyzer, which generates heat during operation (Smolinka et al. 2015). On both sides of the electrolyzer, surplus water is recycled and fed to the input water stream. While the oxygen produced merely passes through a condensation trap before being released to environment, the hydrogen must be treated in a more complex purification process. First, a demister removes fine droplets of liquid water. In the second stage, the remaining oxygen is catalytically removed within two deoxodizers. In this step, water is produced, which is subsequently condensed and removed with the aid of a heat exchanger and a demister, respectively (Smolinka et al. 2015). The resulting product is dry hydrogen with a purity of 99.9% (Siemens AG SILYZER 2019) at a pressure of 35 bar and temperature of 333–353 K (Stadtwerk Haßfurt 2019). As the electrolyzer must be operated with direct current, a power conditioning unit is required. This unit consists of a transformer and a rectifier, which adapt the alternating current (Smolinka et al. 2015).

5.3 Methodology

To determine the environmental burdens of a PEMWE plant, a life cycle assessment (LCA) is performed. This LCA was conducted in accordance with the ISO 14040 and 14044 standards, which describe the steps of implementing an LCA (German Institution for Standardization 2006, 2018). In addition to these general standards, the guidance document specific to LCAs on hydrogen producing systems developed by Lozanovski et al. was applied to some extent (Lozanovski et al. 2013).

5.3.1 Goal and Scope Definition

The goal of this study is to quantify the environmental impacts of hydrogen production through PEMWE. To date, a few studies have already carried out a PEMWE life cycle assessment, as can be seen in a review from Koj et al. (2019). Most of these studies are focused on assessing gaseous and liquid fuels based on hydrogen such as methane or methanol (Koj et al. 2019). Nevertheless, other studies, like those performed by Bareiß et al. (2019) and Wulf and Kaltschmitt (2018), concentrate on the production of hydrogen, which results in a more detailed description of the PEMWE plants that are analyzed. Bareiß et al. as well as Wulf and Kaltschmitt state that the electricity supply is the biggest contributor of environmental burdens and that the impact of producing the hydrogen production plant is negligible (Bareiß et al. 2019; Wulf and Kaltschmitt 2018). Therefore, the aim of this study is to verify these results. The target audience is researchers in the field of synthetic fuels and interested third parties.

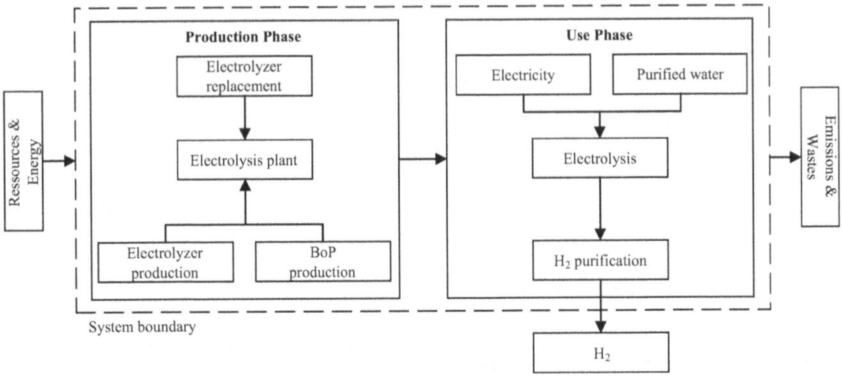

Fig. 5.3 System boundary of the electrolysis plant

As Fig. 5.3 shows, the system boundary of PEMWE envelopes both the production and use of a plant.

The electrolysis plant contains the components illustrated in Figs. 5.1 and 5.2. Since it is expected that more than one electrolyzer is required throughout the lifetime of the BOP, which amounts to 20 years, the production phase also includes the manufacturing of replacement electrolyzer stacks. In contrast, the use phase involves the steps required to be able to produce purified hydrogen. Water is split when purified water and electricity enter the electrolyzer. The evolved hydrogen then must be purified, and this, too, is accounted for in the use phase. The resulting product is purified hydrogen and, thus, the functional unit is defined to be "1 kg of hydrogen with 99.9% purity at 35 bar and 333–353 K". In the following assessment, this functional unit is abbreviated to "kg H_2". The installation of the individual components on the plant site as well as the end of life phase of the product system is disregarded. The electrolysis plant is designed for an electrical input power of 1.25 MW, and the system efficiency amounts to 60% with regards to the lower heating value of hydrogen. It is assumed that the plant is produced in Europe and operated in Germany. Although water electrolysis also produces oxygen, no allocation is applied because it is not utilized but released to the environment.

5.3.2 *Life Cycle Inventory*

Foreground data for the electrolyzer is based on literature. In total, the plant contains five stacks with 50 cells each. The active surface area of each cell is assumed to be 1400 cm^2 (Bessarabov and Millet 2018) and this, in combination with the thickness of the individual cell components (Fig. 5.1), and the respective material densities were used to calculate the mass of the various materials. As PEM electrolyzers have structures similar to those of PEM fuel cells (Ito 2015), suitable processes were adopted from Notter et al.'s life cycle assessment of the latter (Notter et al. 2015).

Foreground data for the BOP materials is based on estimations derived from a real electrolysis plant and validated by the plant operator. In modeling the background system, both the ecoinvent v3.5 database and the GaBi professional database 8.7 were used. The model was built in GaBi ts (version: 9.2).

5.3.3 Life Cycle Impact Assessment

Within the life cycle impact assessment, environmental impacts are assigned to the material and energy flows recorded in the life cycle inventory phase. For this step the life cycle impact assessment method CML 2001 using characterization factors from 2016 (van Oers 2016) was applied. This method contains the following 12 baseline impact categories: depletion of abiotic resources (ADP elements and ADP fossil), acidification (AP), eutrophication (EP), freshwater aquatic ecotoxicity (FAETP inf.), marine aquatic ecotoxicity (MAETP inf.), global warming with and without biogenic carbon (GWP100 and GWP100 excl. bio), human toxicity (HTP inf.), ozone depletion (ODP), photochemical ozone creation (POCP), and terrestric ecotoxicity (TETP inf.). Of these impact categories, the global warming potential including biogenic carbon caused by hydrogen produced through PEMWE is analyzed in greater detail since it is the most commonly discussed topic at present. For the sake of simplicity, in this paper, the global warming potential including biogenic carbon is shortened to "GWP100".

5.3.4 Scenarios

The production of hydrogen requires substantial amounts of electrical energy, and the electricity sector is continuously changing. Thus, the environmental impacts of hydrogen production are analyzed considering three scenarios, which differ in the electricity generation structure as well as in the plant operation mode. The first scenario ("Current") represents the current electricity generation structure in Germany as it is modelled in the GaBi professional database 8.7. Here, it is assumed that the electrolysis plant is continuously operated and, thus, the number of full load hours, after deducting maintenance, amounts to 8000 h per year. Furthermore, the electrolyzer unit is expected to last 40,000 h of operation (Matute et al. 2019), which equals 5 years. In the second scenario, the electricity supply is exclusively based on renewable sources, with 70% originating from wind and 30% from photovoltaics (PV). As these technologies only intermittently provide electricity, the full load hours of the plant are reduced to 4000 h per year. Another consequence of the intermittent supply of electricity is that the electrolysis has to be operated dynamically. Dynamic operation and repeated cold starts may reduce the stack lifetime (Matute et al. 2019). For this reason, the calendaric stack lifetime of 5 years is assumed, which leads to 20,000 operation hours per electrolyzer unit in the scenario using electricity from wind and PV. This scenario is shortened to "Wind + PV". Finally, the electrolysis plant's operation is also considered assuming an electricity supply exclusively stemming from wind power ("Wind"). However, due to the fluctuating supply of electricity from wind power, the amount of full load hours is reduced to 2000 h per year. As a result, it is also assumed that the electrolyzer's lifetime remains 5 years, which results in 10,000 operation hours before it is entirely replaced.

5.4 Results and Discussion

The normalized results of the different impact categories, illustrated in Fig. 5.4, facilitate the direct comparison of the three scenarios regarding their environmental benefits. Throughout the normalization, for each impact category, the scenario with the highest impact is determined and set as reference value for the respective category. In doing so, the proportions of the impacts caused by the scenarios can be displayed separately for each category.

The application of renewable energy is beneficial in 8 of 12 impact categories, regardless of whether the electricity originates from wind and PV—at the specified ratio—or exclusively from wind. Considering categories like ADP fossil, GWP100, and GWP100 (excl. bio), the impacts are reduced by more than 90% due to the plant operation with electricity from renewable energy. Furthermore, the impacts of hydrogen production on AP and POCP decrease by around 77–86%. The categories EP, MAETP inf., and TETP inf. show a different trend. In these categories, the environmental impacts still decrease compared to the current situation by using electricity originating from wind and PV or exclusively from wind. However, the comparison of the two scenarios considering the renewable electricity generation structures reveals

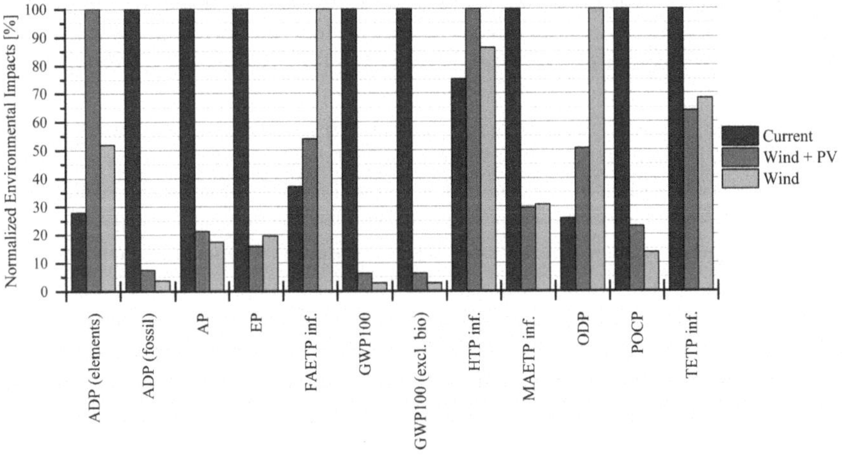

Fig. 5.4 Normalized impact profile of a PEMWE distinguished by the three scenarios

that the electricity supply exclusively stemming from wind power causes slightly higher impacts on EP, MAETP inf., and TEPT inf. than using electricity stemming from wind and PV.

Yet, there are also impact categories that are negatively affected by the use of electricity from renewable sources. With respect to ADP elements, the scenario with mixed renewables has the highest impact, whereas the Wind scenario also leads to an increased depletion of elements compared to the Current scenario. The same relation between the three scenarios is observed for the impact category HTP inf. Thus, with regards to the ADP elements and the HTP inf. assigned to the hydrogen produced, it is more beneficial to use the currently available electricity generation techniques instead of renewables.

Continuing, in considering FAETP inf. and ODP, the environmental burdens of hydrogen produced with PEMWE increase significantly due to the plant operation with renewable energy. While the impact on FAETP inf. increases by a factor of 2.7, the environmental burdens affecting the ODP are almost quadrupled. This is because the production of the plant contributes the largest share to the overall emissions considering these two impact categories. A reduction of the stack lifetime due to dynamic operation resulting from the application of electricity from renewable energy sources, thus, leads to significantly increased impacts on FAETP inf. and ODP. This increase cannot be offset by altered emissions throughout the operation phase.

In contrast, other impact categories, such as ADP fossil, AP, and GWP100 (incl. and excl. biogenic carbon), are mainly affected by the operation phase of the plant at present. This implies that developing the electricity generation structure towards renewable energy may significantly decrease the environmental burdens caused by the production of hydrogen. However, in the following, only the global warming potential including biogenic carbon (GWP100) is analyzed further.

As can be seen in Fig. 5.5, in the Current scenario, the impact of producing the electrolysis plant accounts for 0.2% of the overall GWP100 and, thus, is negligible. The remaining 99.8% of the climate-changing emissions are caused throughout the

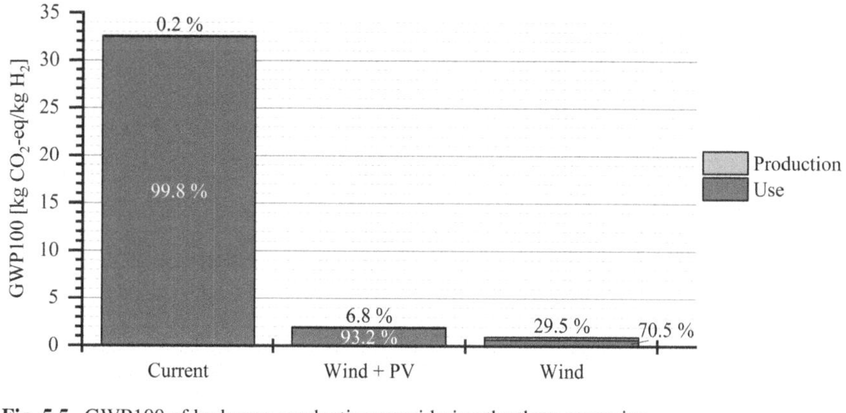

Fig. 5.5 GWP100 of hydrogen production considering the three scenarios

Table 5.1 Values for the GWP100 of hydrogen produced with PEMWE and the reference process steam methane reforming (SMR)

Absolute amount	GWP100 (kg CO_2-eq/kg H_2)	Reduction
Current scenario	32.52	Reference value
Wind + PV	2.02	93.8%
Wind	0.93	97.1%
SMR	10.60	–

operation phase and are mainly attributed to the electricity demand (99.7%) of the plant. The absolute values of greenhouse gases emitted are listed in Table 5.1.

If the electricity required for the operation of the plant is based solely on renewable energy, with 70% from wind power and 30% from PV, the GWP100 of one kilogram of hydrogen substantially decreases (see Table 5.1). For this reason, the production phase relatively grows to 6.8% of the overall emissions. This growth is intensified by the lowered stack lifetime as this leads to less product per stack, to which the emissions can be assigned to. Meanwhile, in the Wind scenario, the GWP100 of one kilogram of hydrogen further diminishes, whereas the contribution of the plant production increases. Consequently, in this scenario, 29.5% of the greenhouse gases per kilogram of hydrogen are emitted during the production of the plant, representing a significant contribution to the overall emissions. As the electricity generation structure develops towards renewable energy and the impact of the operation phase on the GWP100 of hydrogen from PEMWE decreases, the manufacturing must be considered in more detail. For this reason, in Fig. 5.6, the individual components of a PEMWE plant and their effects on the GWP100 are illustrated for the

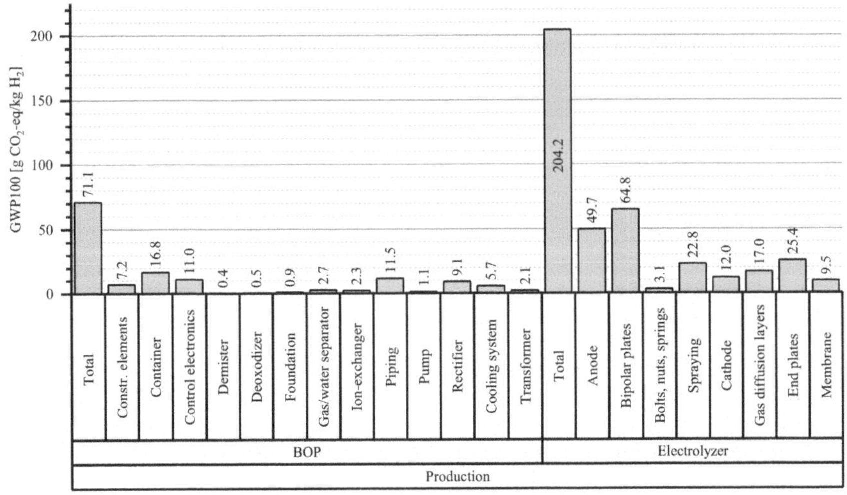

Fig. 5.6 Impact of the production of individual electrolysis plant components on the GWP100 exemplified by the Wind scenario

Wind scenario. Whereas the absolute emissions caused throughout the production of the plant do not differ among the three scenarios, the emissions related to the functional unit vary among the scenarios due to the different lifetimes assumed and, thus, different amounts of product produced per stack. As the Wind scenario has the lowest stack lifetime and, therefore, the lowest product output, the relative emissions of the plant's production phase are higher than those of the Current or the Wind + PV scenario.

The BOP system covers 13 main components, of which the production of the container, the piping, and the power electronics represent the largest emitters. BOP production emits only 71.1 g CO_2-eq/kg H_2, whereas the electrolyzer production releases nearly three times as much greenhouse gas to the environment. It consists of seven main components and is, with regards to the number of components and the mass, significantly smaller than the BOP system. Nevertheless, the electrolyzer's materials and processing steps are associated with higher environmental burdens. The components with the highest impact on the GWP100 are the bipolar plates, followed by the anodes and end plates. Furthermore, the application of the electrocatalysts to the membranes, which is best described as a spraying process, has a substantial effect on the GWP100.

As this study shows, the production of the plant may have substantial impacts on the environmental burdens of hydrogen produced with PEMWE. For this reason, this phase should be analyzed in more detail. Unfortunately, availability of resilient data is limited. While, for the components of the electrolyzer itself, data on the materials and their respective mass is accessible in literature, data on the surrounding BOP components relies solely on estimations and are based on a specific plant. As such, it cannot be universally transferred to other plant sizes, configurations, and locations, for example.

Finally, the impact on the GWP100 is compared to that caused by the most commonly used hydrogen production technology, namely steam methane reforming (SMR) (Dagdougui et al. 2018). A GaBi professional (8.7) dataset reveals the impact on the GWP100 to be 10.60 kg CO_2-eq per kilogram of hydrogen produced. It should be noted that, on the one hand, the SMR process is a well-established process and, on the other hand, the dataset does not represent a single plant, but average emissions of a variety of plants. Nevertheless, the amount of greenhouse gases emitted by SMR represents a threshold for those caused by hydrogen production through PEMWE, indicating when it may environmentally be more beneficial than SMR. In particular, with this value, the maximum GWP100 of the electricity used in the operation phase can be calculated since the GWP100 of the production phase is known. Assuming an amount of 8000 full load hours per year and a stack lifetime of 40,000 h, PEMWE is favorable to SMR if the greenhouse gases emitted per kilowatt-hour do not exceed 189.5 g CO_2-eq. Currently, in accordance with the GaBi professional database 8.7, electricity generated in Germany causes climate-changing emissions amounting to 583.9 g CO_2-eq/kWh_{el}.

However, government policy dictates that the generation structure must shift towards renewable energy, which will lead to decreased greenhouse gas emissions. According to the German energy objectives, in 2030, renewable energy sources should account for 65% of all electricity generation (CDU et al. 2018). Based on this target, the German Institute for Economic Research (DIW) performed simulations assuming an electricity generation structure for 2030. Following this study, in 2030, more than 20% of the electricity still originate from hard coal and lignite (Oei et al. 2019). Modeling this electricity generation structure with the respective GaBi datasets results in the GWP100 of 331.9 g CO_2-eq/kWh$_{el}$, of which a large part is caused by electricity from lignite and hard coal (232.6 g CO_2-eq/kWh$_{el}$). Thus, even in 2030, the greenhouse gas emissions of electricity significantly exceed the 189.5 g CO_2-eq/kWh$_{el}$. This means that, in 2030, the production of hydrogen with PEMWE using conventional electricity, which is partly originating from hard coal and lignite, has a higher impact on climate than SMR. In contrast, an accelerated coal phase out compensated with an increased expansion of renewable energy (85%) would lead to a more favorable electricity generation structure in 2030. The resulting electricity generation structure does not make use of hard coal and lignite anymore, but still uses other fossil energy carriers, like oil and natural gas (Oei et al. 2019). In this scenario, the GWP100 would decrease to 110.6 g CO_2-eq/kWh$_{el}$ and, therefore, the production of hydrogen with a PEMWE would be less climate changing than SMR.

5.5 Conclusion

Currently, the production of hydrogen with PEMWE produces 32.52 kg CO_2-eq/kg H_2. Analyzing the results in more detail reveals that, based on the current electricity generation structure, the operation phase accounts for 99.8% of the climate changing emissions, whereas the production of the plant is negligible. Since the GWP100 of the operation phase is almost exclusively a result of generating electricity, the impact of the operation phase will diminish with the expansion of renewable energy in the next decades. For example, using electricity originating from 70% wind and 30% PV sources leads to the GWP100 of 2.02 kg CO_2-eq per kilogram of hydrogen produced with PEMWE. Provided that the electrolysis plant is operated in a wind-only scenario, the greenhouse gas emissions only amount to 0.93 kg CO_2-eq/kg H_2. In this case, the operation phase still accounts for 70.5% of the GWP100. Yet, with 29.5% of the greenhouse gas emissions resulting from the production of the plant, this phase must also be thoroughly considered. Consequently, due to the development of the electricity sector towards renewable energy, the relative impact of plant production on the GWP100 increases and cannot be disregarded. Concerning the production of the electrolyzer, the bipolar plates, the anodes, the end plates, and the application of the electrocatalysts to the membranes cause the highest shares to the greenhouse gas emissions. Compared to the electrolyzer, the production of the BOP components results in relatively low emissions affecting GWP100.

Apart from the GWP100, other impact categories are analyzed in this study. The results show that the development of the electricity generation structure towards renewable energy is beneficial with regards to most of them, however, the categories

ADP elements, FAETP inf., HTP inf., and ODP represent exceptions since the corresponding impacts increase with use of electricity originating from wind and PV or from wind exclusively.

References

Bareiß K, de La Rua C, Möckl M et al (2019) Life cycle assessment of hydrogen from proton exchange membrane water electrolysis in future energy systems. Appl Energy 237:862–872. https://doi.org/10.1016/j.apenergy.2019.01.001

Bessarabov D, Millet P (2018) PEM Water Electrolysis, vol 1. Hydrogen and fuel cells primer series. Academic Press an imprint of Elsevier, London, United Kingdom

Bessarabov D, Millet P (2018) PEM water electrolysis, vol 2. Hydrogen and fuel cells primer series. Academic Press an imprint of Elsevier, London, United Kingdom

CDU, CSU, SPD (2018) Ein neuer Aufbruch für Europa - Eine neue Dynamik für Deutschland - Ein neuer Zusammenhalt für unser Land: Koalitionsvertrag zwischen CDU, CSU und SPD: 19. Legislaturperiode. https://www.bundesregierung.de/resource/blob/656734/847984/5b8bc23590d4cb2892b31c987ad672b7/2018-03-14-koalitionsvertrag-data.pdf?download=1. Accessed 08 Oct 2019

Dagdougui H, Ouammi A, Bersani C et al (2018) Hydrogen infrastructure for energy applications: production, storage, distribution and safety. Academic Press, London

Fuel Cells and Hydrogen Joint Undertaking (2015) Commercialisation of energy storage in Europe: a fact-based analysis of the implications of projected development of the European electric power system towards 2030 and beyond for the role and commercial viability of energy storage: final report. https://www.fch.europa.eu/sites/default/files/CommercializationofEnergyStorageFinal_3.pdf. Accessed 09 Oct 2019

German Institution for Standardization (2006) DIN EN ISO 14040: environmental management—life cycle assessment—principles and framework (ISO14040:2006). German and English version EN ISO 14040:2006

German Institution for Standardization (2018) DIN EN ISO 14044: Environmental management—life cycle assessment—requirements and guidelines (ISO 14044:2006 + AMD 1:2017). German version EN ISO 14044:2006 + A1:2018

Guy B, Schmidt J, de Villafranca Casas MJ et al (2019) Realizing the promise of Paris: roadmap to a safer climate. https://newclimate.org/wp-content/uploads/2019/01/Paris-Roadmap.pdf. Accessed 16 Oct 2019

IPCC (2019) Summary for policymakers. In: Pörtner H-O, Roberts DC, Masson-Delmotte V et al (eds) IPCC special report on the Ocean and cryosphere in a changing climate, pp 1–42

Ito H (2015a) Current Collectors (GDLs) and Materials. In: Bessarabov D, Wang H, Li H et al (eds) PEM electrolysis for hydrogen production: principles and applications. CRC Press, s.l., pp 147–156

Ito H (2015b) Membranes. In: Bessarabov D, Wang H, Li H et al (eds) PEM electrolysis for hydrogen production: principles and applications. CRC Press, s.l., pp 119–133

Koj JC, Wulf C, Zapp P (2019) Environmental impacts of power-to-X systems—a review of technological and methodological choices in life cycle assessments. Renew Sustain Energy Rev 112:865–879. https://doi.org/10.1016/j.rser.2019.06.029

Lozanovski A, Schuller O, Faltenbacher M et al (2013) Guidance document for performing LCA on hydrogen production systems. Fraunhofer Verlag, Stuttgart

Matute G, Yusta JM, Correas LC (2019) Techno-economic modelling of water electrolysers in the range of several MW to provide grid services while generating hydrogen for different applications: a case study in Spain applied to mobility with FCEVs. Int J Hydrogen Energy 44(33):17431–17442. https://doi.org/10.1016/j.ijhydene.2019.05.092

Millet P (2015) Characterization tools for Polymer Electrolyte Membrane (PEM) water electrolyzers. In: Bessarabov D, Wang H, Li H et al (eds) PEM electrolysis for hydrogen production: principles and applications. CRC Press, s.l., pp 179–217

Notter DA, Kouravelou K, Karachalios T et al (2015) Life cycle assessment of PEM FC applications: electric mobility and μ-CHP. Energy Environ Sci 8(7):1969–1985. https://doi.org/10.1039/c5ee01082a

Oei P-Y, Göke L, Kemfert C et al. (2019) Erneuerbare Energien als Schlüssel für das Erreichen der Klimaschutzziele im Stromsektor: Politikberatung kompakt. https://www.diw.de/documents/publikationen/73/diw_01.c.616181.de/diwkompakt_2019-133.pdf. Accessed 09 Oct 2019

Siemens AG (2019) SILYZER200: High-pressure efficiency in the megawatt range. https://assets.new.siemens.com/siemens/assets/api/uuid:8063d38e3aca74d2a0bcc307fcef435ced002103/version:1524044774/silyzer200-broschure-en.pdf. Accessed 29 Oct 2019

Smolinka T, Ojong ET, Garche J (2015) Hydrogen production from renewable energies-electrolyzer technologies. In: Moseley PT, Garche J (eds) Electrochemical energy storage for renewable sources and grid balancing. Elsevier, pp 103–128

Smolinka T, Ojong ET, Lickert T (2015b) Fundamentals of PEM water electrolysis. In: Bessarabov D, Wang H, Li H et al (eds) PEM electrolysis for hydrogen production: principles and applications. CRC Press, s.l., pp 11–33

Stadtwerk Haßfurt (2019) Power-to-gas anlage. https://www.stwhas.de/stadtwerk/projekte/power-to-gas/. Accessed 29 Oct 2019

Staffell I, Scamman D, Velazquez Abad A et al (2019) The role of hydrogen and fuel cells in the global energy system. Energy Environ Sci 12(2):463–491. https://doi.org/10.1039/c8ee01157e

Stolten D, Krieg D (2011) Alkaline electrolysis—introduction and overview. In: Stolten D (ed) Hydrogen and fuel cells: fundamentals, technologies and applications, 2. Reprint. Wiley-VCH, Weinheim, pp 243–270

United Nations Framework Convention on Climate Change (2015) Paris agreement. https://unfccc.int/sites/default/files/english_paris_agreement.pdf. Accessed 09 Oct 2019

van Oers L (2016) CML-IA database: characterisation and normalisation factors for midpoint impact category indicators. Version 4.8, Aug 2016. http://www.cml.leiden.edu/software/data-cmlia.html. Accessed 26 Sept 2019

Varone A, Ferrari M (2015) Power to liquid and power to gas: an option for the German Energiewende. Renew Sustain Energy Rev 45:207–218. https://doi.org/10.1016/j.rser.2015.01.049

Wang C (2015) Bipolar plates and plate materials. In: Bessarabov D, Wang H, Li H et al (eds) PEM electrolysis for hydrogen production: principles and applications. CRC Press, s.l., pp 135–146

Wulf C, Kaltschmitt M (2018) Hydrogen supply chains for mobility-environmental and economic assessment. Sustainability 10(6):1699. https://doi.org/10.3390/su10061699

Zerrahn A, Schill W-P, Kemfert C (2018) On the economics of electrical storage for variable renewable energy sources. Eur Econ Rev 108:259–279. https://doi.org/10.1016/j.euroecorev.2018.07.004

Chapter 6
Remanufacturing of Energy Using Products Makes Sense Only When Technology Is Mature

Introducing a Circular Economy Indicator for Remanufacturing Based on a Parameterized LCA and LCC Assessment of a Circulation Pump

Torsten Hummen and **Elena Wege**

Abstract In the current debate about Circular Economy (CE) many circularity measurements focus solely on material cycles and resource-efficiency. Hence, remanufacturing is per se seen as a sustainable End-of-Life (EoL) pathway. As CE-indicators do often not take the use-phase into account, that assumption is not necessarily true and the benefit questionable. This is why in this paper a remanufacturing indicator is introduced which evaluates if remanufacturing is a good EoL strategy for energy using products (EuP) from an ecologic and economic point of view, taking technical improvements and energy efficiency gains into account. We elaborate what matters when assessing remanufacturing, especially when a comparative assessment is conducted. The indicator is applied in a case study where remanufacturing of an old circulation pump is compared to using various new, more efficient pumps as substitutes in a 2nd Life Cycle. Thereby, a parameterized model is developed for the LCA and LCC assessment of the remanufacturing process. The main finding is that remanufacturing of EuP only makes sense when technology is mature. This is very distinct from an environmental perspective but applies also to the financial aspects.

Keywords Remanufacturing · Circular Economy · Substitute · Efficiency development · Mature technology · Indicator

---

T. Hummen (✉)
Bosch Thermotechnology GmbH, Junkersstraße 20, 73249 Wernau (Neckar), Germany
e-mail: torsten.hummen@de.bosch.com

Chair for Ecological Systems Design, Institute of Environmental Engineering, ETH Zurich, John-von-Neumann-Weg 9, Zurich 8093, Switzerland

E. Wege
Robert Bosch GmbH, Corporate Sector Research and Advance Engineering, Robert-Bosch-Campus 1, 71272 Renningen, Germany
e-mail: Elena.Wege@de.bosch.com

S. Albrecht et al. (eds.), *Progress in Life Cycle Assessment 2019*,
Sustainable Production, Life Cycle Engineering and Management,
https://doi.org/10.1007/978-3-030-50519-6_6

6.1 Introduction: Remanufacturing as Circular Economy Scheme

The concept of Circular Economy (CE) is experiencing increased attention (Bocken et al. 2017). Even though there is no consensus yet on the definition of CE (Kirchherr et al. 2017), the objective to ensure sustainable consumption is clear in the EU (EC 2015). In the center of CE is the aim to close the material loop and to use materials as long as possible. Extending a products' lifetime with remanufacturing is one scheme for this (Skene 2018). Remanufacturing is the process of restoring a used product to at least like-new condition (BSI 2009). Hence, it is a recovery strategy to close the material loop on low entropy level in order to minimize the employment of exergy and the associated environmental impact (Rechberger and Graedel 2002).

Resource extraction and processing cause half of the total global warming potential (GWP) and more than 90% of land- and water-related impacts (IRP 2019). Respectively, huge monetary flows are still associated with the extraction and procession of raw materials (BGR 2017). Reducing the impact of the raw material phase and thus the resource related environmental impact as well as costs is one aim of remanufacturing. It is achieved by disassembling the old core product and inspecting all parts. Those which pass the inspection are cleaned and reused. Sometimes surface treatment or mechanical processing are involved in remanufacturing as well. Parts, which do not pass the inspection, must be replaced by newly manufactured ones. In the end, the final product is assembled and the lifetime of the core product is extended (IRP 2018).

One of the main research challenges is to determine when to apply which product life extension strategy (remanufacturing, refurbishment, repair and direct reuse) (Bakker et al. 2014). Especially for heating systems and its components as Energy using Products (EuP) this hasn't been done yet because current legislation like the Eco-design Directive in the EU focuses mainly on the use phase (EU 2009) to achieve the greatest leverage in terms of environmental improvements with the least life cycle costs (VHK 2011). A proper environmentally compatible design for the End of Life (EoL) is neglected so far, though the EoL treatment - meaning the collection and further pre-processing - is regulated under the open scope definition of the WEEE Directive (EU 2012).

The analysis of potential environmental impacts of a product or system is usually carried out with a Life Cycle Assessment (LCA), which we also use in our approach. Comparatively, the evaluation of the lowest monetary expenses that come with a product or system throughout the life cycle is done with Life Cycle Costing (LCC).

Several studies are combining the economic and ecologic assessment of a product in order to have a broader decision basis, e.g. (Bierer et al. 2015; Simões et al. 2016). Only few studies so far have focused on combining the LCC and LCA approach considering the remanufacturing of products, e.g. (Boustani et al. 2010; Schau et al. 2011). To the best of our knowledge, none of these have taken a closer look at the remanufacturing process itself, which may be a bias towards remanufacturing (Boustani et al. 2010). This is why this paper aims at closing this gap: A closer look at the remanufacturing is taken and the environmental and economic effects of a remanufactured circulation pump compared to a new one are evaluated, taking energy-efficiency gains and expenses for remanufacturing into account. We try to provide the needed support for facts about the environmental and financial benefits of remanufacturing (Peters 2016) herewith.

The aim of this study is to deal with the question if remanufacturing of EuPs like circulation pumps is a good Design for Environment (DfE) strategy, especially when considering future efficiency developments. We systematically answer this question with LCA and LCC, discuss how to define the right substitute and assess the burden from remanufacturing with a parameterized model to determine when to apply remanufacturing as a DfE strategy. We propose an indicator to evaluate the potential of CE from an environmental and economic point of view, which is currently still not sufficiently covered in CE indicators (Elia et al. 2017). The indicator shows when it makes sense to remanufacture and enables to quickly derive recommendations for action.

6.2 What to Compare: Let's Be Clear About Substitution

Due to the advantage of energy and material savings along the whole value chain, remanufacturing is repeatedly claimed to be environmental beneficial by its own nature (Baxter 2019). In environmental assessments it is often only compared to the corresponding manufacturing process of the same product, e.g. (Ortegon et al. 2012; Sutherland et al. 2008). A comprehensive literature review about the current knowledge of circular principles and environmental impact shows that of 73 analyzed studies dealing with environmental assessment of remanufacturing over 90% of them are still only comparing remanufacturing and manufacturing (van Loon 2019). This falls short especially in the case of EuP, where the use phase has a significant impact on environmental burden (Ardente and Mathieux 2014). When including the use phase and efficiency improvements of new product generations, lifetime extension with remanufacturing is not always to favor (Ardente et al. 2018; Baxter 2019). In those cases, replacing the old product with a more efficient one is of advantage. The extra burden from the production phase of the new product is paid off during the use phase and results in a smaller net burden. Hence, the substitute to compare with must be carefully defined when assessing remanufacturing of EuP.

Any Circular Economy strategy will only be large-scale successful if also economic benefits can be achieved for at least one stakeholder in the value stream, otherwise there is no economic incentive for any of the participating stakeholders to deviate from the current *Take-Make-Dispose* economy (Andersen 2007). Hence, when deciding if remanufacturing is a good EoL strategy also the economic burdens that come with a remanufactured product should be compared to the life cycle costs of a new product in order to gain a more complete picture (Schau et al. 2011). Like in environmental assessment, this is taking energy efficiency gains and process yields of remanufacturing into account.

In a linear economy where you *Take, Make, (Use) and Dispose* a product, a product is scrapped at the EoL and materials are recycled. With remanufacturing the use cycle of the product is extended, so that it is used for a second full service life and circulates back in the economy ($Scenario_{Reman}$). The environmental and economic value of the material retains. Here, we assume a simple case with only two full service lives. It means that the product reaches its EoL after the second cycle and must be further processed at this point in time. Complete recycling of the initial product is not necessary, only parts which fail the inspection have to be replaced and hence recycled.

To meet the operation time of two full service lives a second new product can be used instead as well. This product can either be the same as the initial one ($Scenario_{New,\ same}$) or a more mature technology, which is more efficient ($Scenario_{New,\ more\ efficient}$). When using a new product in the second cycle the first initial product must be recycled completely. The same applies for the second product.

The described scenarios differ in terms of the necessary recycling scheme, the new production of (primary) materials and, last but not least, in the use phase. Accordingly, two cases have to be distinguished in a comparative scenario assessment:

- Case A: $Scenario_{Reman}$ compared to $\text{Scenario}_{\text{New, same}}$
- Case B: $Scenario_{Reman}$ compared to $\text{Scenario}_{\text{New,}\ more\ efficient}$

Figure 6.1 shows all phases along the life cycles as symbolic representation and highlights below what life cycle phases need to be considered (color marked boxes) and differ in each case (Case A: T_1, M_2, P_2, Case B: T_1, M_2, P_2, U_2).

We follow the elaboration by (Ardente et al. 2018) and state the impact along the whole life cycles as the sum of impacts from primary material phase (M), production phase (P), use phase (U) and EoL Treatment (T). Moreover, the impacts have to

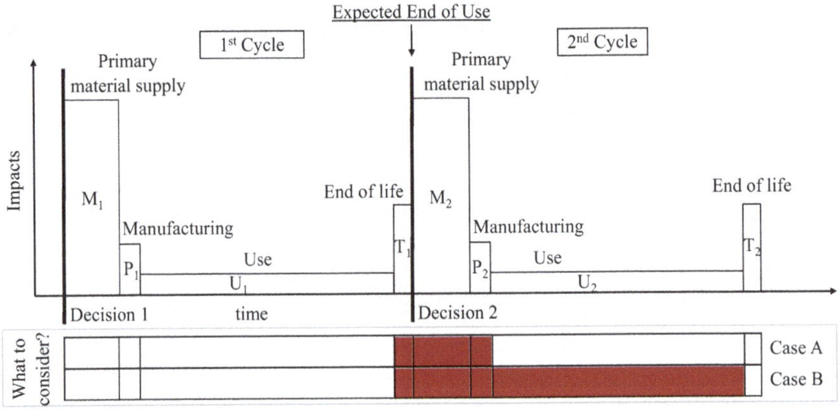

Fig. 6.1 Use Cycle extension with remanufacturing or new product; symbolic presentation based on (Frischknecht 2010); Life cycles which differ when comparing substitutes are displayed for each case with color marked boxes. (Case A: $\text{Scenario}_{\text{Reman}}$ vs. $\text{Scenario}_{\text{New, same}}$; Case B: $\text{Scenario}_{\text{Reman}}$ vs. $\text{Scenario}_{\text{New, more efficient}}$)

be added up according to the number of use cycles (i = number of use cycles, j = materials in product). In the case of two use cycles l is equal to two, which we will only consider further below.

$$I_x = \sum_{i=1}^{l}(\sum_{j=1}^{k}(M_{j,i}) + P_i + U_i + \sum_{j=1}^{k}(T_{j,i})) \tag{6.1}$$

Life cycle phases which equal out in the comparison (see Fig. 6.1, color marked boxes) can be ignored. We assume that all products have the same or similar material composition and that EoL treatment of the second product can be neglected. The same applies to the necessary spare parts requirements during the use phase. Hence, Eq. (6.1) can be simplified for each case and structured so that the avoided impact ($I_{avoided,\ Case\ x}$) is calculated for the comparative assessment. In the remanufacturing scenario ($Scenario_{Reman}$) only materials which do not pass the inspection and must be replaced have to be considered (z = replaced materials), whereas in the new production scenario ($Scenario_{New}$) all materials must be considered. In case A the use phases $U_{2,Reman}$ and $U_{2,Substitute}$ are the same, equal out and can hence be neglected.

$$I_{avoided,Casex} = I_{Substitute} - I_{Reman} \tag{6.2}$$

$$\begin{aligned} I_{avoided,Casex} = & \sum_{j=1}^{k}(T_{j,1})_{Substitute} + \sum_{j=1}^{k}(M_{j,2})_{Substitute} \\ & + P_{2Substitute} + U_{2Substitute} - \sum_{z=1}^{r}(T_{z,1})_{Reman} \\ & - \sum_{z=1}^{r}(M_{z,2})_{Reman} - P_{2Reman} - U_{2Reman} \end{aligned} \tag{6.3}$$

If $I_{avoided}$ is bigger than zero, it means that the impact of the substitution scenario is higher and remanufacturing is preferable. But if $I_{avoided,\ Case\ A} < I_{avoided,\ Case\ B}$ then case 2 should be preferred to the first one. To conclude, all possible available technologies must be considered to make sure that all substitutes are assessed and to ensure that the most beneficial substitute is included in the assessment.

Circular Economy Indicator for remanufacturing of Energy using Products (EuP). Based on the discussion before, a Circular Economy Indicator for remanufacturing showing when it makes sense to remanufacture is necessary. Such an Indicator has not been introduced yet. Most existing CE indicators fall within the resource-efficiency cluster (Parchomenko et al. 2019) and focus on the preservation of material (Moraga et al. 2019). These are for example the MCI (Material Circularity Index) (Ellen MacArthur Foundation 2015), the CEIP (Circular Economy Indicator Prototype) (Cayzer et al. 2017) or the REAPro method (Resource Efficiency Assessment of Product) (Ardente and Mathieux 2014). They have the underlying assumption in common that reuse (including remanufacturing) is beneficial to the environment, which is not per se true. It must be assessed more comprehensively, also in the context of Circular Economy, by sufficient consideration of the use phase with a Life Cycle Thinking (LCT) approach.

The Life Cycle Gap Analysis (LCG-A) (Dieterle et al. 2018) indicates the potential of recycling in the context of Circular economy and the REV (Retained Environmental Value) (Haupt and Hellweg 2019) calculates the best EoL pathway. Both approaches are based on the interpretation of LCA results but do not state when remanufacturing makes sense explicitly. A distance to target approach is not included directly and hence no ranking is applied.

As for LCC, there are a number of key performance indicators that indicate the financial performance of a CE strategy. To the author's concern, most of these, like the Value-based-resource efficiency (VRE), give results on a macro level, not on micro or product level (Di Maio et al. 2017). Other indicators like resource-efficiency indicators do combine a LCA and LCC perspective (Pauliuk 2018), but are only concerning the material phase and not a holistic life cycle perspective. This is why we introduce an indicator which covers both aspects and which gives some indicative guidance on remanufacturing as such. To our knowledge there is currently no other indicator assessing remanufacturing with a relative approach for LCA as well as for LCC.

We propose a 2-stage Circular Economy Remanufacturing Indicator (CERI) and calculate the distance between the avoided impact from one case x to the avoided impact from the case with the highest avoided impact. As avoided impact is the impact that can potentially be saved with a scenario, the higher the avoided impact, the better. This is then normalized by $Reman_{norm}$ which indicates from when on remanufacturing could be a preferable pathway in a scenario assessment:

$$CERI_i = \begin{cases} 0 & , if\, I_{avoided,caseA} < 0 \\ \dfrac{\frac{1}{\left(I_{avoided,best} - I_{avoided,x}\right)}}{Reman_{norm}} & , if\, I_{avoided,x} < Reman_{th} \\ 1 & , if\, I_{avoided,x} \geq Reman_{th} \end{cases} \tag{6.4}$$

$$Reman_{th} = I_{avoided,best}(1 - u_c) \tag{6.5}$$

$$Reman_{norm} = \frac{1}{\left(I_{avoided,best} - Reman_{th}\right)} \tag{6.6}$$

As the CERI specifies the same for LCA and LCC, only one explanation is given here. Figure 6.2 shows a graphical interpretation of the calculation steps. CERI results are in a range between 0 and 1: The further the avoided impacts are apart, the closer to 0 the indicator is and the less beneficial remanufacturing is. The closer the result is to 1, the more favorable it is. If the avoided impact is above the threshold value $Reman_{th}$ then the CERI is set to 1 (Eq. 6.4, $I_{avoided,\,x} \geq Reman_{th}$). This means that the avoided emissions are high enough that remanufacturing should be realized.

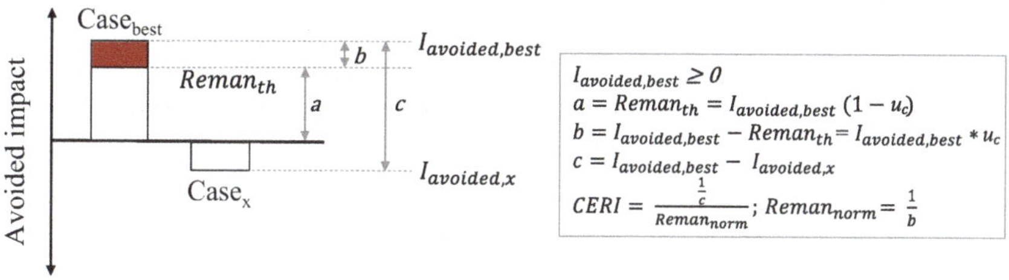

Fig. 6.2 Graphical interpretation of the calculation steps for the CERI

Otherwise the formula from Eq. 6.4 for $I_{avoided,\,x} < Reman_{th}$ must be applied. When remanufacturing has a higher impact than new production, CERI is set to 0 (Eq. 6.4, $I_{avoided,\,case\,A} < 0$). We base our formula on a number of assumptions:

1. Uncertainty u_c (in %). A general uncertainty is assumed. Values for the uncertainty can vary, depending on scope and assumptions of the study. This is due to data quality and general uncertainty that occurs whenever a Life Cycle (Cost) Assessment is done. By including uncertainty in our formula and not just in a sensitivity analysis we hope to increase acceptance for the CERI among potential users.
2. The threshold $Reman_{th}$ indicates, at which point the distance between the various options is so small that remanufacturing could be applied. Inherent uncertainty of a life cycle analysis is met with this approach. The smaller the assumed uncertainty, the closer the threshold is to the best-case result.

An advantage of the method is that allocation, like in the circular footprint formula (EC 2018) for example, to the specific products in each use cycle is not necessary. We are addressing the meaningfulness of remanufacturing in total, not just of a singular product, so that the need for credits of avoided burden is eluded. Further, our previous elaboration shows that the analysis of the first life cycle is not necessary except for the EoL-phase. This relative approach also significantly decreases the amount of required data for a comprehensive analysis.

The applicability of the CERI is shown with a case study on remanufacturing of a circulation pump. We apply the CERI indicator to various exemplary impact categories, which are GWP as proxy for several midpoint LCA indicators and $ReCiPe_{Endpoint,\,total}$ as single score endpoint indicator. Moreover, we calculate the indicator for the costs.

6.3 Case Study on a Circulation Pump

A circulation pump is the main electricity-using component in gas boilers and is used to circulate the heating fluid to the radiators. Like every electrical motor, it consists

Table 6.1 Inventory data of a circulation pump

Material composition[1]		Assembly[2] and distribution[3]	
Steel	1.607 kg	Electricity	1.742 kWh
Copper	0.831 kg	Heat	11.449 MJ
Aluminum	0.188 kg	Water	1.3E-05 m^3
Polyamide	0.103 kg		
Composite Material (PA66 GF30)	0.069 kg	Distribution	216 km
Ceramic	0.020 kg		
Rubber	0.007 kg		

Data are simplified and materials/processes making small contributions to overall impact are consolidated or omitted. (1: Own measurement, 2: scaling according to (Auerbach et al. 2017), 3: according to case study)

Table 6.2 Energy demand of reference circulation pumps per year for analyzed scenarios (with 6840 h working hours) (Topp et al. 2010); EEI: Energy Efficiency Index

Energy efficiency class	C (0.6 < EEI < 0.8)	B (0.4 < EEI < 0.6)	A (0.3 < EEI < 0.4)	A* (0.2 < EEI < 0.3)
Electricity demand (kWh/year)	287.14	217.72	171.00	76.60

of a rotor with magnets and stator with copper wirings. The material composition is listed in Table 6.1.

Circulation pumps are grouped into energy efficiency classes based on the Energy Efficiency Index (EEI) (Topp et al. 2010). The EEI indicates how much energy is needed relative to a reference product. The energy demand of reference circulators, which are used as scenarios in our case study, is listed per EEI respectively per Energy Efficiency Class in Table 6.2. A low-efficiency pump with an Energy Efficiency Class "C" runs 285 days a year, which results in an energy demand of 287 kWh per year ($P_{rated} = 44$ W). As the efficiency of circulation pumps has drastically improved in recent years the needed rated power today is at a level of 19 W and the energy demand is 77 kWh per year.

Based on a case study of a remanufactured low-efficiency circulation pump (Energy efficiency class: C), general process rules for remanufacturing are derived (e.g. which parts have to be replaced) in a first step. Secondly, impacts from the remanufacturing process itself are scaled to the system of a circulation pump based on a reference product. This tiered approach leads to a parametrized LCA model for remanufacturing. The environmental impact of different systems is then assessed in a scenario analysis. The same is done for the LCC analysis, where the financial impacts of different replacement scenarios are evaluated.

We assume that circulation pumps of different efficiencies have the same material composition, although more efficient pumps do most likely have other electronic components and probably more copper content. However, the material composition does not vary much and does not have a big impact on the assessment, so a modified material composition is neglected for simplicity reasons. Accordingly, all pumps have the same environmental impact in the production phase and have the same price of approximately 149.00 € (Hornbach 2019). The same applies for remanufacturing as well where rubber parts (e.g. sealings) and standard screws must be replaced. All other parts must be cleaned, repaired and can be reused after successful inspection. A process yield of 80% is assumed, meaning that of every pump which is coming back to the remanufacturing process 80% are functioning and can be remanufactured, whereas 20% of parts have to be recycled.

The geographical scope of the study is Germany. This holds especially for the life cycle stages remanufacturing, assembly, use and EoL. It furthermore implies that the pumps are used in Germany and that broken pumps are transported back for remanufacturing to a facility in Göttingen, Germany. Moreover, we use the German electricity mix in our assessment.

The design time of a circulation pump is 15 years, which is also the time frame for the full service life in each use cycle considered here. As the electricity mix can change quite a lot over this time span, we project the future electricity mix for Germany with the respective environmental impact based on (Prognos 2014). Corresponding price developments for electricity have been calculated based on IHS Data (IHS 2013).

Recycling is assessed with the WEEE treatment dataset from Ecoinvent to calculate the environmental burden from EoL treatment. This applies to the 20% of pump parts which cannot be remanufactured ($\text{Scenario}_{\text{Reman}}$) or to the complete pump respectively, in case of a new pump ($\text{Sceanrio}_{\text{New, same}}$ or $\text{Scenario}_{\text{New, more efficient}}$). For costs, scrapping revenue for old E-Machines has been used as calculation basis because the material composition is similar.

As GWP have been demonstrated to correlate with ozone depletion, acidification and eutrophication for resource extraction and processing (Steinmann et al. 2016), results of the study are given in GWP as midpoint indicator. It is also the most relevant impact category for the use phase of EuP. Thereby we do not want to neglect the impact of other impact categories but rather use GWP as a proxy for several categories. In addition to that, we use the $\text{ReCiPe}_{\text{Endpoint, total}}$ indicator as a single score impact category. The environmental assessment is conducted with Ecoinvent (version 3.4 cutoff system model) as background database (Wernet et al. 2016).

For the Life Cycle Costing, expenses are the only impact category. Because LCA is a burden based approach, the economic impact is also expressed as financial burden (Carlsson 2005). Main assumptions are that no price changes in manufacturing technology and price differences between product generations have been assessed. Discounting has been calculated with an inflation rate of 1.5% and a discount rate of 5.0% (Fleischner 2000). Other costs and price developments have been calculated with expert data, industry averages and internal data. In addition, in different countries there might be government subsidies for buying a circulator with an energy efficiency level of A* or better. However, this cannot be addressed here properly as it depends very much on the user-situation and is therewith out of the scope of the study. Taxes have also been excluded from this study as they do not change the monetary relation between manufacturer and user.

The scenario results of our case study are interpreted with the introduced CERI. For that, we assume a general uncertainty u_c of 10% to calculate the mentioned threshold $Reman_{th}$.

6.3.1 Parameterized Model for Remanufacturing Assessment

To the authors' best knowledge, impact from remanufacturing is often neglected or modelled as a fraction of the initial product manufacturing. For example, in (Boustani et al. 2010) the energy required for the remanufacturing process is zero or in (Baxter 2019) a fraction for the remanufacturing is defined, but then assumed to be negligible. To overcome this problem we developed a parameterized model for the remanufacturing assessment.

In prospective LCA missing data are often assumed with analogous processes or systems (Arvidsson et al. 2013) and scaled to the system under consideration (Hummen and Kästner 2014). The reference product and the system under consideration must be similar. We follow the scaling approach by (Shibasaki 2008) here and estimate the impact of remanufacturing based on a reference study of an E-Machine with detailed information about involved processes (Auerbach et al. 2017). From the reference, data about energy and auxiliary masses are obtained. This allows us to take into account scaling effects for each individual process and to adopt the impact to the system under consideration. However, contrary to (Shibasaki 2008), we neglect scaling and learning effects when merging all process steps. A process yield is nevertheless incorporated.

Based on our case study we identified reverse logistics and cleaning as well as the final inspection at the end of the manufacturing line as main contributors to the environmental impact. This is followed by Disassembly as well as Re-Assembly and Recycling. Especially the assembly steps are mainly manual processes and hence have a small environmental impact when assessing remanufacturing from an environmental perspective. However, they have significant impact on the costs, as manual assembly steps are very cost-intensive. We assume here that the process of Re-assembly is the same as Disassembly, which again is the same as when producing the initial or more mature product.

The scaling factors of the parameterized remanufacturing model for all assembly processes are based on an interface ranking with α_k values according to the approach by CEN/CLC (2017) as well as the product weight. Product volume is used as scaling parameter for the cleaning process and process yield (incl. products that fail entrance exam and inspection on part level) for recycling. The reverse logistic is assumed with distribution distance of the case study and End of Line inspection is neglected, as it is the same in all assessed cases.

A workshop on remanufacturing of components used in heating systems was conducted to derive general rules for replacement of parts and process yield. The workshop was carried out internally at Bosch with experts for the product and the remanufacturing process.

6.3.2 *Results*

In case of a circulation pump, the remanufacturing process is causing 84% less GWP respectively even 98% less $\text{ReCiPe}_{\text{Endpoint, total}}$ damages than new production of the same pump. However, the biggest environmental impact is still caused during the use phase, as according to (Prognos 2014) the assumed electricity mix in Germany has still a high share of coal and gas power (over 50%) over the entire use phase of 15 years. The conducted scenario analysis shows that a remanufactured low-efficiency circulation pump only has a better environmental performance over all life cycles if the substitute has an efficiency increase by not more than 0.06 W when considering GWP respectively 0.73 W for $\text{ReCiPe}_{\text{Endpoint, total}}$. This is not the case if a potential substitute is a more mature and efficient pump. Then use cycle extension with a new and more mature circulation pump is environmentally beneficial. The ideal LCA-case would be upgrading an old pump of an energy efficiency class lower than A* to A*.This is represented in Fig. 6.3 by the CERI results. Remanufacturing circulation pumps with an energy efficiency class of (C – A*) are compared to a newly produced pump with an energy efficiency class of A*. If the result is smaller

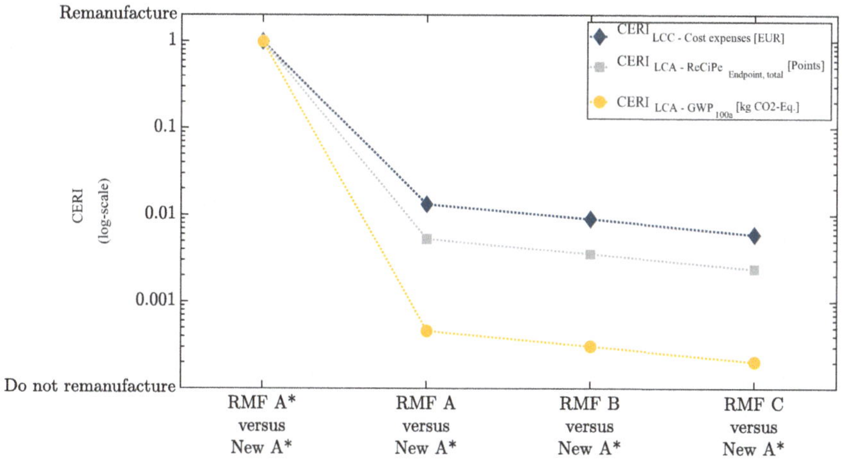

Fig. 6.3 Circular Economy Remanufacturing Indicator (CERI) results for remanufactured circulation pumps (RMF) compared to a new and more mature circulation pump (New) with an efficiency class of A*. 1 indicates the best case scenario. The higher the value, the more favorable remanufacturing. It shows the environmental ($\text{CERI}_{\text{LCA - GWP, 100a}}$, $\text{CERI}_{\text{LCA - ReCiPe, Endpoint, total}}$) and cost ($\text{CERI}_{\text{LCC}}$) results when remanufacturing pumps with different energy efficiency classes of C – A* (see Table 6.2 for efficiency classification)

than one it means that remanufacturing of the corresponding efficiency class does not make sense and use cycle extension with the new pump is environmentally favorable.

The low $CERI_{LCA}$ values for scenarios with a remanufactured circulation pump lower than efficiency class A* imply that remanufacturing is not environmentally beneficial. That is due to the significant impact caused during the use phase for all energy efficiency classes: over 99% when considering GWP and over 93% for $ReCiPe_{Endpoint,\ total}$. Remanufacturing should therefore be avoided, except if an initial pump with a low energy efficiency class can be upgraded to A*. In that case, the $CERI_{LCA}$ would be one.

Results for LCC are similar to LCA results. Although costs for remanufacturing of the circulation pump are lower than for new production, the comparatively high costs of the use phase compensate all the potential savings of remanufacturing, as overall much more cost can be saved with a more efficient circulation pump. Due to the predominant impact of the use phase, the most favorable cost situation is the replacement of an old pump with a new pump that is at least one energy efficiency class better—showing that even a small efficiency increase has quite a significant effect. This is especially the case for pumps being installed now, as electricity prices in Germany have risen significantly in the last decade (IHS 2013). The ideal LCC-case would be upgrading an old pump of an energy efficiency class lower than A* to A*, as here most costs can be saved. However, this is not always a feasible option. This aspect is discussed later on. Our approach shows that although the use phase has a significant impact on Life Cycle Costs, the impact of remanufacturing itself cannot be neglected. This will be the case especially for less use-phase and more material phase intensive products.

$CERI_{LCC}$ results are slightly better than from an environmental perspective, shown in Fig. 6.3. However, remanufacturing is still not to be recommended regarding costs in this assessment if the circulation pump has not the best energy efficiency class of A*.

6.4 Discussion

Circular Economy is often understood to transform the *Take, Make, (Use) and Dispose* economy to a more sustainable one. However, this is not true by default as we analyzed for the value retention process of remanufacturing. The results show that remanufacturing only makes sense when a technology is mature and high efficiency gains are not possible anymore. This is well reflected by the CERI. If technology is mature and differences between product generations become smaller, the range of indicator results will decrease. However, in certain situations replacement of an old product with a new generation is not always possible. This could e.g. be the case when a circulation pump needs to be replaced in an existing system where a new pump does not fit into the periphery. In this case remanufacturing is still to be the preferred option, because it still has a lower environmental and financial burden than a newly produced pump of the old generation.

Albeit future electricity mixes are already considered, results will change significantly for the environmental assessment if the used electricity mix is less fossil intensive. In our assessment, the environmental impact from electricity decreases by 6% for GWP and 5% for $ReCiPe_{Endpoint, total}$ over the full service life of 15 years. With a more renewable electricity mix the use phase is less significant and remanufacturing could then make sense environmentally. However, those influencing parameters can be reflected by the CERI with further scenarios as impacts are interpreted and normalized to a scale of 0 to 1.

Here we only considered remanufacturing to restore the like-new condition without the option of upgrading. This has a big influence. Upgrading even one energy efficiency class is reducing environmental impact as well as financial burden of a 2nd life cycle when compared to an old remanufactured pump that is not upgraded. When compared to a new pump, the upgrade needs to be in the same range as the most mature technology, otherwise upgrading is not beneficial ecologically or economically. This is also shown in the CERI results. However, there are some technical limitations to the upgradability of a product. For example, there might be new requirements for interfaces that are not compatible with the former design.

There are other methodological limitations of our approach. This concerns especially intertemporal aspects. Whereas intertemporal effects are addressed in LCC via a discount rate, they are only considered in the characterization factors but are otherwise not assessed with the LCA logic yet (Bierer et al. 2015).

The scaling factors of the parametrized model for LCA and LCC of remanufacturing are based on a similar product (E-Machine) and focus on interfaces and weight of a product. These parameters have proven to be crucial for the dismantling-ability and remanufacturing process but need further validation in some case studies.

6.5 Conclusion

When assessing remanufacturing of EuP the potential substitute must be well defined. Whether a substitute has the same efficiency or can be upgraded to a higher efficiency class has a big influence on the system to be preferred. Neglecting potential efficiency gains in the use phase might lead to a wrong conclusion as a replacement with a better technology is not considered and thus underestimated. When the use phase is causing the biggest environmental impact and is allocated to the biggest financial burden as well, newer product generations must always be considered. As conclusion, one can say in general that remanufacturing makes sense when an energy-using technology is mature.

By introducing the indicator approach CERI we want to give practical guidance on answering the question if remanufacturing makes sense for an EuP or if another strategy should be pursued. With our relative indicator not all life cycle phases must be evaluated but only those that actually need to be considered when comparing a new product to a remanufactured old one. Effort for data collection and conducting LCA and LCC can be reduced significantly.

The CERI interprets and normalizes LCA and LCC results so that recommendations for action can be derived quickly from a scenario analysis. It is acting as a supplement rather than a substitution for many other assessments like the LCG-A or REV and can be integrated into these. Moreover, the indicator can be applied to other EuPs, but also to non EuPs whose main environmental and financial burden lies in the material-, assembly or EoL-phase. Then even fewer life cycle stages need to be considered.

Strictly following the waste hierarchy and avoiding waste with circular economy principles like remanufacturing is not always ideal. Circular Economy is not burden free. It is neglected by material based indicators used to measure circular economy and can only be mitigated with more comprehensive LCT approaches and indicators, especially in the case of EuP where the use phase must be in the scope of the assessment.

References

Andersen Mikael Skou (2007) An introductory note on the environmental economics of the circular economy. Sustain Sci 2(1):133–140

Ardente F, Mathieux F (2014) Identification and assessment of product's measures to improve resource efficiency: the case-study of an Energy using Product. J Cleaner Prod 83:126–141

Ardente F, Talens Peiró L, Mathieux F, Polverini D (2018) Accounting for the environmental benefits of remanufactured products: method and application. J Cleaner Prod 198:1545–1558

Ardente F, Mathieux F (2014) Environmental assessment of the durability of energy-using products: method and application. J Cleaner Prod 74:62–73

Arvidsson R, Kushnir D, Sandén BA, Molander S (2013) How to make policy-relevant life cycle assessments of future products? Lessons learned from nanomaterials. In Proceedings of the 6th international conference on life cycle management, pp 72–75

Auerbach R, Brämer T, Brouwer E, Dierks C, Gassmann A, Dörr M, dos Santos C, Miehe R, Öhl J, Schmid K, Seikel E, Wüst H (20170 RECVAL-HPM. Innovative RE-use and ReCycling VALue Chain for High-Power Magnets. Schlussbericht

Bakker C, Wang F, Huisman J, den Hollander M (2014) Products that go round: exploring product life extension through design. J Cleaner Prod 69:10–16

Baxter J (2019) Systematic environmental assessment of end-of-life pathways for domestic refrigerators. J Cleaner Prod 208:612–620

Bundesanstalt für Geowissenschaften und Rohstoffe (BGR) (2017) Deutschland – Rohstoffsituation 2016, Hannover

Bierer A, Götze U, Meynerts L, Sygulla R (2015) Integrating life cycle costing and life cycle assessment using extended material flow cost accounting. J Cleaner Prod 108:1289–1301

Bocken NMP, Olivetti EA, Cullen JM, Potting J, Lifset R (2017) Taking the circularity to the next level: a special issue on the circular economy. J Industrial Ecol 21(3):476–482

Boustani A, Sahni S, Graves SC, Gutowski TG (2010) Appliance remanufacturing and life cycle energy and economic savings. In: 2010 IEEE international symposium on sustainable systems and technology. IEEE, pp 1–6. https://doi.org/10.1109/ISSST.2010.5507713

BSI (2009) Design for Manufacture, Assembly, Disassembly and End-of-life processing (MADE). Terms and definitions. British Standards Institution, 8887-2

Carlsson RM (2005) Economic assessment of municipal waste management systems - case studies using a combination of life cycle assessment (LCA) and life cycle costing (LCC). J Cleaner Prod 13(3):253–263

Cayzer S, Griffiths P, Beghetto V (2017) Design of indicators for measuring product performance in the circular economy. Int J Sustain Eng 10(4–5):289–298
CEN/CLC (2017) General methods for the assessment of the ability to repair, reuse and upgrade energy related products. prEN 45554. Accessed 24 May 2019
Di Maio F, Rem PC, Baldé K, Polder M (2017) Measuring resource efficiency and circular economy: a market value approach. Resour Conserv Recycling 122:163–171
Dieterle M, Schäfer P, Viere T (2018) Life cycle gaps: interpreting LCA results with a circular economy mindset. Proc CIRP 69:764–768
Elia V, Gnoni MG, Tornese F (2017) Measuring circular economy strategies through index methods. A critical analysis. J Cleaner Prod 142:2741–2751
Ellen MacArthur Foundation (2015) Circularity indicators. An approach to measuring circularity. Methodology
European Commission (EC) (2015) Closing the loop. An EU action plan for the circular economy
European Union (EU) (2009) Directive 2009/125/EC of the European Parliament and of the Council of 21 October 2009. Establishing a Framework for the Setting of Ecodesign Requirements for Energy-related Products
European Union (EU) (2012) Directive 2012/19/EU of the European Parliament and of the Council of 4 July 2012 on Waste Electrical and Electronic Equipment (WEEE)
European Commission (EC) (2018) Product Environmental Footprint Category Rules Guidance. Version 6.3. European Commission: Brussels, Belgium
Fleischner G (2000) Eco-Design. Effiziente Entwicklung nachhaltiger Produkte mit euroMat. Springer, Heidelberg
Frischknecht R (2010) LCI modelling approaches applied on recycling of materials in view of environmental sustainability, risk perception and eco-efficiency. Int J Life Cycle Assess 15(7):666–671
Haupt M, Hellweg S (2019) Measuring the environmental sustainability of a circular economy. Environ Sustain Indicators 1–2:100005
Hornbach (2019) Heizungspumpen & Umwälzpumpen bei HORNBACH kaufen. https://www.hornbach.de/shop/Heizungspumpen-Zubehoer/Heizungspumpen/S773/Marke=Grundfos/artikelliste.html. Accessed 10 Oct 2019
Hummen T, Kästner F (2014) Evaluating existing methodological approaches for prospective LCA at early technology-development stages. In: Proceedings from the LCA XIV International, pp 25–36
IHS (2013) Herausforderungen an die deutsche Wettbewerbsfähigkeit durch einen neuen Weltenergiemarkt, Berlin
International Resource Panel (IRP) (2018) Re-defining Value: The Manufacturing Revolution. Remanufacturing, Refurbishment, Repair and Direct Reuse in the Circular Economy. A Report of the International Resource Panel. Nairobi, Kenya
International Resource Panel (IRP) (2019) Assessing global resource use: a systems approach to resource efficiency and pollution reduction
Kirchherr J, Reike D, Hekkert M (2017) Conceptualizing the circular economy: an analysis of 114 definitions. Resources Conserv Recycling 127:221–232
Moraga G, Huysveld S, Mathieux F, Blengini GA, Alaerts L, van Acker K, de Meester S, Dewulf J (2019) Circular economy indicators: What do they measure? Resources Conserv Recycling 146:452–461
Ortegon K, Nies LF, Sutherland JW (2012) Remanufacturing: an alternative for end of use of wind turbines. Leveraging Technology for a Sustainable World. Springer, Heidelberg, pp 155–160
Parchomenko A, Nelen D, Gillabel J, Rechberger H (2019) Measuring the circular economy - a multiple correspondence analysis of 63 metrics. J Cleaner Prod 210:200–216
Pauliuk S (2018) Critical appraisal of the circular economy standard BS 8001:2017 and a dashboard of quantitative system indicators for its implementation in organizations. Resources Conserv Recycling 129:81–92

Peters K (2016) Methodological issues in life cycle assessment for remanufactured products: a critical review of existing studies and an illustrative case study. J Cleaner Prod 126:21–37
Prognos (2014) Entwicklung der Energiemärkte - Energiereferenzprognose. Prognos AG, EWI, GWS (57/12)
Rechberger H, Graedel TE (2002) The contemporary European copper cycle: statistical entropy analysis. Ecol Econ 42:59–72
Schau EM, Traverso M, Lehmann A, Finkbeiner M (2011) Life cycle costing in sustainability assessment - a case study of remanufactured alternators. Sustainability 3(11):2268–2288
Shibasaki M (2008) Methode zur Prognose der Ökobilanz einer Großanlage auf Basis einer Pilotanlage in der Verfahrenstechnik. Ein Beitrag zur Ganzheitlichen Bilanzierung. Dissertation, Universität Stuttgart
Simões CL, Figueirêdo de Sá R, Ribeiro CJ, Bernardo P, Pontes AJ, Bernardo CA (2016) Environmental and economic performance of a car component: assessing new materials, processes and designs. J Cleaner Prod 118:105–117
Skene KR (2018) Circles, spirals, pyramids and cubes: why the circular economy cannot work. Sustain Sci 13(2):479–492
Steinmann ZJN, Schipper AM, Hauck M, Huijbregts MAJ (2016) How many environmental impact indicators are needed in the evaluation of product life cycles? Environ Sci Technol 50(7):3913–3919
Sutherland JW, Adler DP, Haapala KR, Kumar V (2008) A comparison of manufacturing and remanufacturing energy intensities with application to diesel engine production. CIRP Ann 57(1):5–8
Topp T, Labanca, Nicola (2010) Energy + Pumps Project. Training module for installation contractors. Manual for the trainer
van Loon P (2019) Circular principles and environmental impact reductions: current knowledge and the way forward. LCM2019, Posen
van Holsteijn en Kemna BV (VHK) (2011) Methodology for Ecodesign of Energy-related Products MEErP 2011. Final Report. https://ec.europa.eu/docsroom/documents/26525 (accessed May 2019)
Wernet G, Bauer C, Steubing B, Reinhard J, Moreno-Ruiz E, Weidema B (2016) The ecoinvent database version 3 (part I): overview and methodology. Int J Life Cycle Assess 21(9):1218–1230

Chapter 7
Integrating Environmental Assessment of Emerging Materials into the Material Selection Process

Malte Schäfer, Felipe Cerdas, and Christoph Herrmann

Abstract Existing material selection methodologies do not account for effects that scaling has on the resulting environmental consequences. The impact of producing an emerging material using a lab-scale process is likely to differ significantly from the impact of mass-producing the same material at industrial scale. We propose a methodology, based on the established Ashby material selection process, that combines material selection with environmental impact assessment and takes into account the effects of scaling up material production processes for emerging materials. The approach adds additional steps for data collection, life cycle assessment and scaling to the pre-existing steps in Ashby's methodology. It allows for the assessment of multiple environmental impact categories as well as a transparent way of handling underlying data uncertainty and variability. We review existing approaches of environmental material selection methodologies, derive requirements for a novel approach, present our concept methodology and demonstrate it using a case study example. Further refinement of the concept should primarily address the life cycle assessment and scaling steps.

Keywords Material selection · Material development · Product development · Emerging materials · Novel materials · Scaling · Life cycle assessment · LCA · Environmental impact assessment

7.1 Introduction

In many cases, the materials used to make a product contribute significantly to the environmental impact of that product. Thus, the nature and amount of materials required to make a product often determine its environmental impact (Ashby 2012;

M. Schäfer (✉) · F. Cerdas · C. Herrmann
Chair of Sustainable Manufacturing and Life Cycle Engineering, Institute of Machine Tools and Production Technology (IWF), Technische Universität Braunschweig, Langer Kamp 19b, 38106 Brunswick, Germany
e-mail: malte.schaefer@tu-braunschweig.de

S. Albrecht et al. (eds.), *Progress in Life Cycle Assessment 2019*, Sustainable Production, Life Cycle Engineering and Management,
https://doi.org/10.1007/978-3-030-50519-6_7

Allwood et al. 2011). Material selection processes within product development therefore have far-reaching consequences for product sustainability. Common requirements in material selection address technical (e.g. tensile strength) and economic (e.g. material cost) aspects. For these aspects, and for established materials, methods and tools supporting decision-making during the material selection process already exist (Ashby 2012). However, for users seeking to include emerging materials (e.g. novel, bio-based polymers and composites) in the selection process and to extend the requirements to environmental criteria, there is currently no suitable methodology available. In addition, existing environmental assessments often do not consider the life cycle perspective. Furthermore, the impact assessment is usually limited to carbon emissions and does not take other types of environmental impacts into account.

The fact that most methods do not take into account scale-up effects of material production constitute another obstacle towards the uptake of an emerging material. Considering the case of a large manufacturer, a significant demand increase for an emerging material may have consequences, such as the deployment of the material at an economies-of-scale level. As a result, the environmental impact (and cost) of a material may change. Existing methods, however, generally rely exclusively on current, non-scaled material data.

An early stage assessment may help to avoid developing and using materials that will have undesired environmental consequences. The described interdependency between material selection and material production, and the consequences on environmental impact, is currently under-addressed. Thus, we find there is a need to identify materials that will contribute towards the reduction of environmental impacts of products, and to select them in a systematic matter in order to accelerate the development and market acceptance of those materials.

We propose a methodology that allows product and material developers to assess the environmental impact of emerging and frontier materials in comparison to established materials. The environmental impact shall be assessed along the whole product life cycle, with a focus on the raw material and production stage. The methodology shall take into account interactions between material selection and material production, in such a way that potential future impacts of a scaled-up material production process are assessed using suitable scaling methods.

This paper is structured as follows: an analysis of current available methodologies and identify a research gap is presented in Sect. 7.2. Based on this gap, we outline the steps of the methodology in Sect. 7.3. In Sect. 7.4, we demonstrate its application using a case study example. Finally, we conclude with a summary and outline of future research need in Sect. 7.5.

7.2 Consideration of Environmental Aspects in Current Material Selection Strategies

The material selection process during product development has the purpose of identifying a single candidate material or a list of candidate materials that best fulfil the requirements stemming from the intended application. In order to enable users to assess to what extent a material fulfils a requirement, a material selection process needs to contain a step in which the requirements are stated and described. The requirements may encompass technical, economic and environmental aspects (the latter are sometimes expanded to social criteria as well). Furthermore, the process has to include a step in which materials are evaluated with respect to how their characteristics and properties contribute to meeting the stated requirements. Finally, some kind of ranking shall follow from the evaluation, to allow for the identification of one or several highest ranked material(s).

When considering environmental impacts of materials, the processes involved in producing, using, and disposing of these materials become important factors. Studies on material selection based on environmental criteria consider this by including requirements that refer to the inputs and outputs of these material life cycle processes.

In the following sub-section, we describe the standard material selection process according to Ashby, to illustrate the individual steps in more detail.

7.2.1 Ashby Material Selection Procedure

The Ashby material selection procedure is comprehensive, as it relies on proper ontologies for materials, material families and material properties. It takes into account a large number of relevant requirements for and properties of materials, and it provides a multitude of analytical and visual decision support tools. The approach has been highly influential, which is reflected in the number of citations of related studies, and the number of researchers employing and refining the underlying methods (e.g. Qiu et al. 2013; Broeren et al. 2016). In addition, the approach has been developed and improved upon for more than two decades, and is the subject of multiple books (Ashby 2012; Ashby & Johnson 2013). Thus, we render the quality, validity and relevance of the Ashby material selection process to be high. In the following paragraphs, we briefly describe the procedure (see Fig. 7.1) both in general terms and with a focus on environmental assessment.

An optional first step (**Eco-audit**) when selecting materials based on environmental criteria is to identify the life cycle stage that matters most matters most (Ashby 2012). This decision is usually based on the embodied energy, or greenhouse gas (GHG) emissions related to a material, as these are the two indicators available in the CES Selector software tool, which is based on the Ashby approach (Ashby et al. 2009). Then, the user is to limit the material selection to those materials that

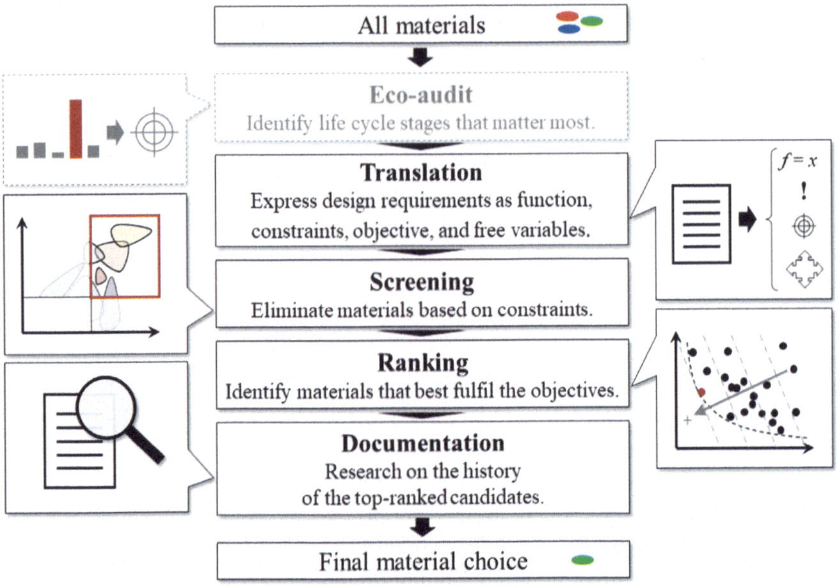

Fig. 7.1 Ashby material selection procedure, inspired by Ashby (2012)

are suited to minimize environmental impact during that stage by defining suitable objectives.

In the **translation** step, design requirements are translated into functions, constraints, objectives and free variables. They describe:

- Function: What does the component do? (e.g. support a load)
- Constraint: What non-negotiable conditions must be met? (e.g. maximum density)
- Objective: What is to be maximized or minimized? (e.g. embodied energy)
- Free variable: What parameters of the problem is the designer free to change? (e.g. color).

Starting with a database of potential materials, constraints are defined in **the screening** step and serve as "gates" through which materials have to pass in an elimination process. In the CES Selector, a large database of materials from which to choose from is included.

During **ranking**, the "surviving" materials are ranked based on objectives. An objective, in nearly all cases, describes an indicator that should be maximized or minimized. A Pareto frontier is a useful visualization to handle trade-offs between two objectives. These can be handled using penalty functions. A penalty function describes the utility of one objective in terms of the other.

The last step of the process consists of seeking out **documentation** for the short list of materials found to be the best solutions. If a penalty function was used, the short list may contain only one candidate. Relevant documentation include case studies of previous uses of the material, failure analyses, reports on its behavior under various conditions, information on its availability, pricing, environmental impact (incl. toxicity and recycling) (Ashby 2012). If a material has been used previously in similar applications that would be considered a positive indicator.

A more design-focused (and less engineering-driven) strategy than the one just described, can be found in (Ashby & Johnson 2013). The following issues remain unchanged, however.

7.2.2 Relevant Aspects Not Included in the Ashby Procedure

While the material selection procedure by Ashby is widely used, adapted and refined, we argue that it lacks some functionalities regarding the environmental assessment – especially with respect to emerging materials.

The materials database taken into consideration during the screening process may not contain a certain material if it is relatively new (e.g. currently only produced at a lab-scale).[1] Even if it does exist in the database, the data point may reflect only its current (or former, outdated) properties, which may be subject to rapid change (e.g. if the material production process is scaled up). Finally, even for well-established materials such as aluminum, the data varies widely (Ashby 2012), due to differences in production processes, measurement procedures, data quality, assumptions, modeling choices and other factors. However, the visualization techniques used in the selection procedure may convey a false sense of certainty regarding the values describing material properties, including their environmental impact.

The issue of data precision is discussed in Ashby (2012), but is not well-reflected in the overall material selection process. Conducting the material selection process with uncertain data may cause a misled selection due to undeserved elimination or survival of materials during the screening stage. Furthermore, solutions identified during the ranking stage may be sub-optimal, as some data points may be misplaced, missing or conveying a false sense of certainty. Finally, documentation will be incomplete or non-existent for novel materials.

Furthermore, the only environmental aspects typically considered in the Ashby procedure are embodied energy and/or GHG emissions, information on the toxicity of substances and material recycling rates. Other types of environmental indicators are not included. This may leave the user with blind spots to environmental impacts such as acidification, eutrophication and particulate matter formation, all of which are categories potentially covered by a life cycle assessment (LCA). The selection of impact categories is part of the goal and scope definition of each LCA study. Assessing

[1] While Ashby does discuss novel materials (Ashby & Johnson 2013), the strategy is much more qualitative and less systematic than the material selection procedure described above. For a quantitative environmental assessment of novel materials, it does not seem fit.

multiple impact categories avoids problem shifting between different types of environmental impacts. Many novel, bio-based materials, for example, are linked to intensive land-use and emissions from agricultural processes, which may be overlooked by conventional metrics such as CO_2 emissions and CO_2-equivalent impacts.

In the next section, we review the literature to identify studies which address some or all of the gaps described above.

7.2.3 Literature Review

Sun et al. (2004) propose a simplified method to evaluate the environmental impact during material selection. Its central aspect is a grouping of materials based on the type of material and environmental performance (Eco Indicator 99 points). The number of groups (16) is determined as the point past which only marginal improvements in average deviation occur. The average environmental performance of a group shall replace the detailed analysis of each individual material. Additionally, the most relevant substances and impact categories for each material group (contributing more than 85% to the final Eco Indicator 99 score) are identified. Using a coffee maker as a case study example, the authors demonstrate that the results from using their simplified approach deviate 11–26% from the results of a complete LCA, regarding the material life cycle stage. Taking into account the complete life cycle, the deviation decreases to 2%.

The approach by Ribeiro et al. (2008) combines technical, economic and environmental aspects. It allows the user to weigh these aspects against one another using a ternary diagram. The importance of individual material properties is derived from the weighting of requirements, and the contribution of each material property to a requirement. Economic aspects are evaluated based on a life cycle costing model, environmental aspects refer to Eco Indicator 99 single-score results. The authors demonstrate the method in a case study of an automotive fender, including a sensitivity analysis on material thickness.

Qiu et al. (2013) integrate structural design and material selection into a hybrid optimization model with the goal of reducing environmental impacts. Their approach differs from others by combining these two aspects into one method, thus considering not just mass- or weight specific environmental impacts of a material, but the impact of the final part. The screening of material properties is based on Ashby's procedure (Ashby 2012), while the assessment of environmental impacts relies on the Eco Indicator 99 method. A dowel bar serves as a case study for demonstrating the approach, in which C45 steel is identified as the optimal solution.

Broeren et al. (2016) present a material selection framework for incorporating environmental criteria and a simplified cost analysis into early stage product (re)design, using the case study of replacing fossil-based polymer printer panels with bio-based alternatives. The basis for comparison is the main product function, not material weight or volume. The framework is aimed at incremental product redesign, and is closely related to the Ashby procedure. It includes additional steps for material testing (e.g. to validate material properties of experimental materials) and ranking using multi criteria analysis. The latter aggregates environmental impacts into one score using normalization and weighting. The results include a range of uncertainty, and a sensitivity analysis for different weighting of the three impact categories (non-renewable energy use, greenhouse gas emissions and land use).

Hallstedt and Isaksson (2017) propose a methodology to support material selection in early product design based on material criticality assessment for alloy materials. Criticality refers to sustainability and availability aspects in this context. After identifying potential critical elements in the pool of considered materials, the level of criticality is graded for each alloy using a sustainability compliance index. The final criticality score of a material includes the number of critical elements in an alloy, the mass composition of the alloy, and the number of existing methods for assessing material availability and sustainability that flag an element as critical.

The framework by Kappenthuler and Seeger (2019) is aimed at material developers, and intends to aid in the identification and prioritization of research projects related to the development of materials for sustainable construction. The four assessment categories are durability, economics, sustainability and future availability, with five to nine indicators in each category. The environmental assessment is based on recyclability, renewability and ReCiPe endpoint scores. The steps consist of defining a goal and boundary conditions, defining a timeframe for analyzing future availability, defining category and attribute weights as well as the functional unit for comparing materials, defining criteria for material screening, selecting candidate materials, assigning scores to attributes for all materials, and finally analyzing strengths and weaknesses to identify areas for improvement.

In order to compare the reviewed approaches in a systematic way, we define seven requirements for a new methodology to be used by practitioners. It is primarily aimed at product developers, however, it should also support the communication between product and material developers, by indicating the future potential of emerging materials. This way, material developers receive guidance on which materials they should develop further. The requirements are:

1. The environmental impact assessment shall **cover all life cycle stages** of the material. This includes changes in energy demand during the use stage due to weight changes, and material specific end-of-life routes.
2. The methodology shall **be applicable to multiple material types** (e.g. metals, ceramics, bio-based materials) and application domains (e.g. office furniture, consumer products, packaging).
3. The approach shall take into account that current **material production processes may change** in the future. This is especially the case for emerging materials, which are produced using processes that may change once significant demand increases occur (interdependency of material selection and material development). A suitable approach should be used to project how current production processes may scale in the future.

Table 7.1 Degree to which the reviewed studies meet the selected criteria

References	(1)	(2)	(3)	(4)	(5)	(6)	(7)
Ashby (2012)	●	●	○	○	○	○	●
Sun et al. (2004)	●	●	○	○	◑	◑	○
Ribeiro et al. (2008)	●	●	○	○	◑	◑	●
Qiu et al. (2013)	●	●	○	○	◑	○	●
Broeren et al. (2016)	●	◑	○	○	◑	●	●
Hallstedt and Isaksson (2017)	○	◑	○	○	◑	○	○
Kappenthuler and Seeger (2019)	●	◑	●	●	●	●	○

4. Analogous to requirement #3, it follows that the methodology should take into account the fact that **technical properties of materials may change** in the future, as they are developed further.
5. The impact assessment shall **differentiate between multiple environmental criteria**. This excludes single-score results that cannot be dissected to analyze the contribution of individual impact categories, in order to be able to uncover trade-offs between different kinds of environmental impacts.
6. The methodology shall **address sensitivity, variability and uncertainty**, and how these affect outcomes and results. Ideally, a visual representation should guide the user to handle decision making when confronted with fuzzy results.
7. Finally, the methodology should preferably **build on existing tools and methods**, ideally well-established and widely used ones. This should ease adoption for the user and allows the developers of the method to focus on the most important points that need to be addressed.

7.2.4 Future Research Demand

Table 7.1 summarizes the evaluation of the six studies reviewed above and the Ashby material selection procedure, using the seven criteria we identified to be relevant.

If criterion #2 is only partially fulfilled, the methodology is generally suited to all kinds of materials, but some aspects make it material specific (e.g. evaluating the environmental impact based on land-use, a criterion mostly relevant to bio-based materials). As most approaches use single score results, they receive partial points for criterion #5 if they at least allow for a dissection of different endpoint indicators (e.g. human health, resources and ecosystems). A partial fulfilment of criterion #6 describes the case where at least the standard deviation of eco indicator points from the group norm is presented (Sun et al. 2004), or when a sensitivity analysis regarding the weighting is included (Ribeiro et al. 2008).

All but one approach consider the complete life cycle, and some are only or primarily applicable to certain types of materials (bio-based polymers, alloys or construction materials). Only one study takes into account potential future changes in material properties and production processes. Most methodologies rely on single-score results for assessing environmental impacts, which do not allow a differentiation of different types of impacts. Many methodologies do not touch on the subject of uncertainty and sensitivity of results. Half of the strategies are based on existing tools and methods, in two cases the Ashby procedure for material selection (Ashby 2012; Ashby & Johnson 2013; Ashby et al. 2009).

While the approach by Kappenthuler and Seeger (2019) appears to meet most requirements, it has some drawbacks. It is aimed at construction materials, therefore the criteria that are used reflect requirements for this specific subsection of materials. Also, the method is semi-quantitative, and in some categories qualitative (e.g. for recycling rates). The quantitative categories rely on comparative ranking (percentiles), which does allow for assessment of the absolute environmental impact.

Therefore, we decide to build our methodology on top of the comprehensive, well-established Ashby procedure, and expand it to meet the requirements stated above. The methodology intends to fill the research gap identified in the literature review.

7.3 Towards a Material Selection Methodology for Emerging Materials Based on Environmental Criteria

Having identified no material selection procedure which fulfils all requirements regarding the level of details, underlying uncertainty and variability, as well as the development potential of materials and material production processes in the future, we proceed by describing the concept of a methodology which covers all these aspects (see Fig. 7.2).

The first step is identical to the **translation** step of the material selection procedure as proposed by Ashby. The "Eco-audit" step is skipped, as environmental life cycle aspects are covered later on in the procedure. We assume that developers start with an existing material they want to replace, as well as with an existing application for this material. Starting with the application, the first step consists of extracting all functions, constraints, objectives and free variables for the material (Ashby 2012).

Next, developers need to **collect data** for the novel material(s) to be included in the selection process. This data shall include material data (e.g. physical properties), process data (e.g. process efficiencies, inputs and outputs) and economic data for the screening procedure (e.g. material & energy cost). Potential data sources include primary data from own measurements, material supplier information, scientific and industry literature, expert knowledge and information available on the internet. Some estimates and assumptions will definitely be necessary at this point. For lab-scale processes, the data collection step shall include information on potential future upscaling – i.e. how the processes to produce the material will change if the production output is to grow by a factor of 10, 100 or 1000, and how the respective resource

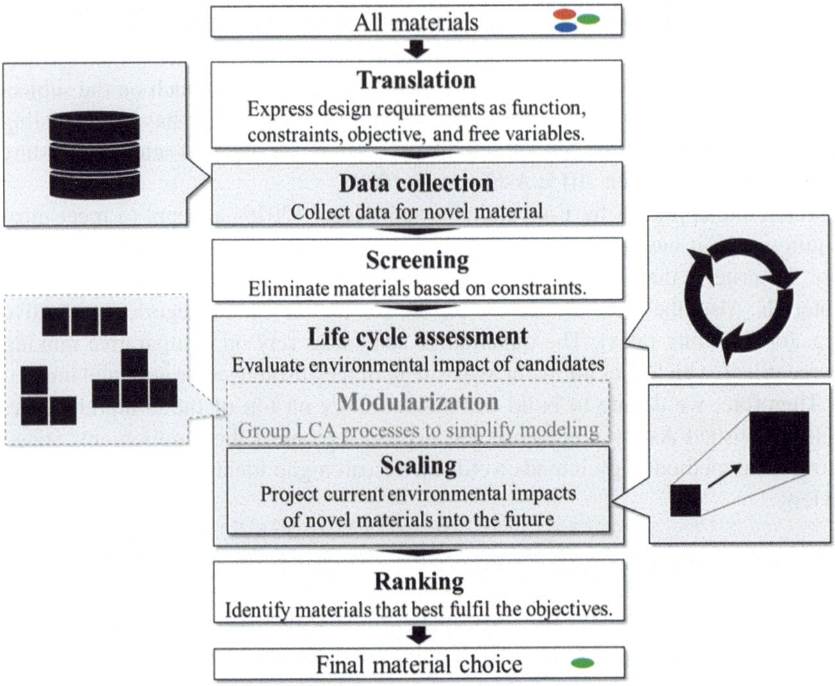

Fig. 7.2 The proposed methodology for material selection based on environmental criteria for novel materials, building on the Ashby material selection procedure

and emission flows change accordingly. The data collection step is largely identical to building a life cycle inventory when conducting a LCA. Therefore, LCA literature may be helpful, e.g. for handling missing, incomplete and low quality data (ISO 2006a; Hauschild et al. 2018).

In a third step, developers conduct the **screening** process based on technical and economic constraints, including the novel materials. For characterizing novel materials, it is essential that all relevant information have been retrieved in the data collection step (e.g. physical characteristics). For characterizing uncertainty and variability for the underlying data, developers should use appropriate ways to describe and communicate both. A pedigree matrix, a semi-quantitative method used in LCA databases to describe data quality, may serve as guidance on which factors may influence accuracy and precision (Frischknecht and Jungbluth 2007; Ciroth et al. 2016).

After having eliminated all but a few materials, developers (or, if available, LCA experts) are then to **conduct a LCA** of the remaining candidates. The ISO guidelines on LCA, the ILCD guidelines and books can provide the necessary background information for this step (ISO 2006a; Hauschild et al. 2018; European-Commission 2011). The implementation will usually require commercial (SimaPro, GaBi or Umberto) or free (Open LCA, Brightway) software as well as access to databases (e.g. ecoinvent,

GaBi, USETox, exiobase). For cases in which information on one or more life cycle stages is sparse (typically the use stage), it is vital to do scenario- and sensitivity analyses, in order to grasp the range of potential results. This task becomes much easier with Python-based, calculation-efficient software packages such as Brightway, as has been demonstrated for the integrated computational life cycle engineering (icLCE) approach using the case study of batteries for electric vehicles (Cerdas et al. 2018).

In order to simplify the LCA of similar materials, the LCI models may be **modularized** (Shibasaki et al. 2007). By grouping together processes that appear in the same combination across several types of materials, the required modeling effort may decrease significantly.

The LCA results then serve as input to the **scaling** method, which translates current environmental impacts stemming from a novel material produced in a lab-scale process into impacts that may occur with an industrial, large-scale material production process (Shibasaki et al. 2007; Piccinno et al. 2016; Arvidsson and Molander 2017). This is done by modifying the life cycle inventory to reflect the changes of the production process during upscaling. It requires process-specific knowledge, which may stem from expert interviews, analyzing upscaled processes of similar materials, the literature and from physical and technical constraints. The outcome of the scaling step is an updated life cycle inventory of the novel material, with input and output flows adjusted accordingly. These result in changes regarding the life cycle impact assessment (LCIA). Both the modularization and the scaling step take place within the LCA. The LCA results should be calculated for both the initial and the scaled-up inventory, in order to allow for a comparison of both cases with reference materials.

The scaled LCIA results, the original (current state) results for the novel material(s), and the results for all conventional materials that survived the screening process, serve as input to the **ranking** of materials. To this end, all technical and economic objectives are plotted pairwise against environmental objectives, e.g. material cost vs. global warming potential (GWP) or tensile strength vs. abiotic resource depletion. The latter consist of all environmental impact categories (e.g. global warming potential GWP, acidification potential AP, abiotic resource depletion potential ADP) agreed upon in the goal and scope definition of the LCA. The resulting plots allow for a visual comparison and decision support for handling trade-offs between technical (or economic) and environmental criteria, as well as handling internal trade-offs among different types of environmental impacts. They include information on uncertainty and variability related to data points, and allow the user to assess the changes resulting from upscaling material production processes.

The final step from the Ashby procedure, **documentation**, is modified to accommodate the circumstances of lab-scale production. Typically, case studies and previous applications of a material won't be available in this case. Instead, the documentation phase should focus on collecting and structuring all available information at this early stage, including information on data quality and assumptions.

In the following section, we demonstrate the methodology using a fictitious case study informed by published LCA results.

7.4 Case Study: Vehicle Seat Cover

In this case study, we assume a practitioner seeks to find an alternative material to bovine leather for car seat covers. After conducting some research, the practitioner finds out that a material made from ground up, dried apple skins mixed with polyurethane (PU) appears to be a promising alternative to the animal based reference material. However, apple leather is still a very novel material, so only little is known about it.

Translating the design requirements of a car seat into functions, constraints, objectives and free variables may result in something similar to Table 7.2.

Some of these requirements such as “quality feel” and “quality look” are quite “soft” and cannot be extracted directly from material databases. Instead, some additional background information is required, such as customer opinions on the “look” and “feel” of the materials in question.

For data collection on leather, the developer may turn to material databases and software such as the Granta CES Selector, published studies on leather, its technical properties and its life cycle environmental impacts (Black et al. 2013) as well as information from the leather industry. For apple leather, the developer could contact companies that currently produce the material at a small scale, review publicly available information on the internet, and compare the production process to that of conventional, synthetic PU leather (Chwang et al. 2002). The data collection process should be driven by the initially formulated functions, constraints and objectives: all these criteria have to be covered by the material and process data. The data quality will be assessed using a pedigree matrix (Frischknecht and Jungbluth 2007; Ciroth et al. 2016).

This case study uses a simplified screening process. Usually it would involve a visual representation of material properties, as depicted for the screening process in Fig. 7.1. Since the case study contains only two materials, it is sufficient to check if both materials meet all constraints. However, a visual representation of the data

Table 7.2 Translation of design requirements for a car seat cover

Function	Constraint	Objective	Free variable
Protect seat cushion	Tensile strength >8 MPa	Minimize cost	Color
Support driver load	Tear force >20 N	Minimize weight	
Repel dirt	Survive >100.000 load cycles (stretching)	Minimize wear	
Transmit heat	Survive >30.000 load cycles (bending)	Minimize abrasion	
Transmit water vapor	Electrical resistivity $>10^8$ μΩ*cm	Minimize environmental impact	
Absorb moisture	“Quality feel”		
	“Quality look”		

points, including underlying uncertainty, may be helpful for comparing both alternatives. Figure 7.3 illustrates how the data quality assessed using a pedigree matrix could correlate with the visual representation of data points for two materials A and B.

The subsequent life cycle assessment of the two types of leather provides information on the respective environmental impacts. For leather, published LCA studies are available (Notarnicola et al. 2011), which can be used to compare the results from one's own LCA model. For apple leather, no LCA studies are available, so the

Indicator score	1	2	3	4	5
Reliability	1.00	1.05	1.10	1.20	1.50
Completeness	1.00	1.02	1.05	1.10	1.20
Temporal correlation	1.00	1.03	1.10	1.20	1.50
Geographical correlation	1.00	1.01	1.02		1.10
Further technological correlation	1.00		1.20	1.50	2.00
Sample size	1.00	1.02	1.05	1.10	1.20

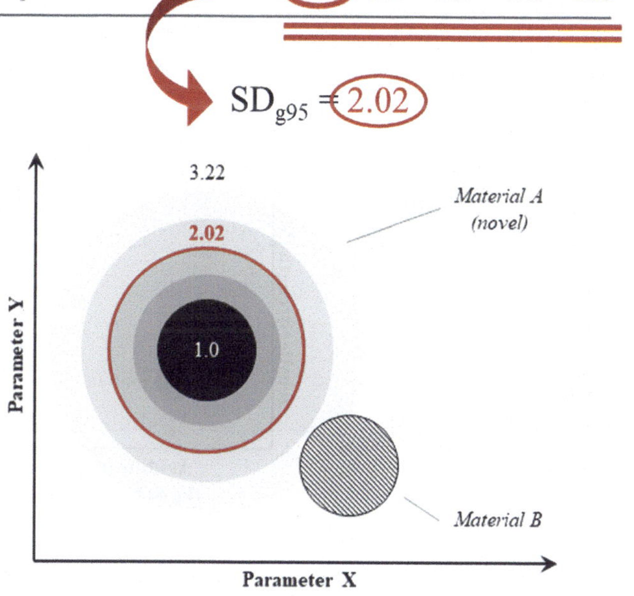

Fig. 7.3 Visual representation of data uncertainty using a pedigree matrix. The total data uncertainty is calculated using an aggregated indicator score (SD_{g95}, the square of the geometric standard deviation, 95% interval). The exemplary indicator score (2.02) is visualized by the red ring around the solid black data point for the novel material A. The maximum indicator score in this case is 3.22 (highest uncertainty), the lowest is 1.0. Material B is an established material with idealized data uncertainty of 1.0. We assume the basic uncertainty factor to be 2.0 for all cases (min/max/example). For more information on the pedigree matrix, and how SD_{g95} is calculated, see Frischknecht and Jungbluth (2007), Ciroth et al. (2016)

LCI model has to be built based on collected data, assumptions, comparisons and information from the literature.

For the scaling step, apple leather is compared to other bio-based and petrochemical materials, which share some of its characteristics. For example, some production steps may be similar to large-scale PU leather production once (and if) the production of apple leather is scaled up. If the machines used for large-scale production are more efficient (energy and material use per m^2 of apple leather produced), the resulting environmental impact may be lower. By adjusting the material and energy flows in the LCI model of the apple leather, according to the scaling estimates based on the analogous polyurethane leather process, the resulting environmental impacts are modified. These impacts represent the case of a hypothetical, future state for a large-scale production of apple leather.

These environmental impacts, expressed as indicators for multiple impact categories, can now be ranked against other technical and economic criteria. By comparing multiple impact categories, environmental trade-offs become apparent. Visualizing underlying uncertainty for the data points avoids instilling a false sense of certainty in the user. In Fig. 7.4, two environmental impact categories are ranked against a technical objective.

While the values for both materials are fictitious, they illustrate (a) trade-offs between different environmental impact categories, (b) the current and future uncertainty regarding the data points for novel materials and (c) how the scale-up potential

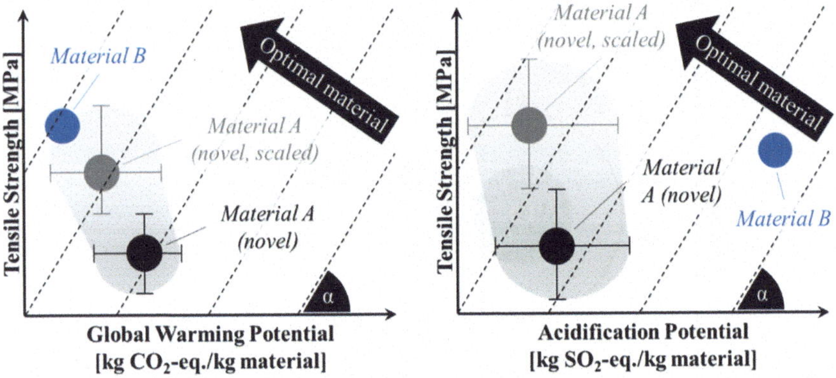

Fig. 7.4 Ranking of two materials A and B (e.g. leather and apple leather—values shown here are fictitious) against one another, using the objectives "maximizing tensile strength" and "minimizing global warming potential" (left) /"tensile strength" and "acidification potential" (right). Here, uncertainty is visualized using error bars for the novel material B, in addition to a shaded area symbolizing the uncertain location of the data point. On the left, material B is likely to dominate material A even for a scaled up production process, while on the right, material A dominates with a high degree of certainty for both cases. α determines the slope of the lines indicating the trade-off between the two objectives. A large α (steep slope) indicates that relatively small increases in global warming/acidification potential have to yield relatively large increases in tensile strength for two materials to be of equal utility (and vice versa), according to the penalty function defined by the user

of materials can be visually communicated. By using the depicted Pareto charts with penalty functions (dashed lines), it is possible to select the material according to individual preferences for environmental, economic and technical criteria (economic criteria are not depicted in Fig. 7.4).

7.5 Summary and Outlook

The proposed methodology builds on the Ashby material selection procedure, while addressing some research gaps regarding the environmental impact assessment, data uncertainty and novel materials. It combines several aspects that no other methodology that we were able to retrieve covers as well.

Of course, several points still need to be addressed. The proposed concept needs further refinement and detailing. The case study is not yet suited to demonstrate the effectiveness of the methodology, and should be followed by a quantitative case study employing the refined approach. A better understanding is needed which scaling approaches are best suited for the case of environmental impact assessment of material production processes, and these approaches need to be implemented in the methodology. It would also benefit from further simplifying the LCA step of the process, as this is usually quite time- and resource consuming. The challenge lies in not losing too much information during the simplification step.

Therefore, our next steps consist of refining the methodology, especially the LCA and scaling steps, and applying it to a real case study.

References

Allwood JM, Ashby MF, Gutowski TG, Worrell E (2011) Material efficiency: a white paper. Resour Conserv Recycl 55(3):362–381. https://doi.org/10.1016/j.resconrec.2010.11.002

Arvidsson R, Molander S (2017) Prospective life cycle assessment of epitaxial graphene production at different manufacturing scales and maturity. J Ind Ecol 21(5):1153–1164. https://doi.org/10.1111/jiec.12526

Ashby MF (2012) Materials and the environment: eco-informed material choice. Elsevier

Ashby MF, Johnson K (2013) Materials and design: the art and science of material selection in product design. Butterworth-Heinemann

Ashby MF, Miller A, Rutter F, Seymour C, Wegst UGK (2009) The CES eco-selector—background reading

Black M, Canova M, Rydin S (2013) Best available techniques (BAT) reference document for the tanning of hides and skins. Eur Comm. https://doi.org/10.2788/13548

Broeren MLM, Molenveld K, van den Oever MJA, Patel MK, Worrell E, Shen L (2016) Early-stage sustainability assessment to assist with material selection: a case study for biobased printer panels. J Clean Prod 135:30–41. https://doi.org/10.1016/j.jclepro.2016.05.159

Cerdas F, Thiede S, Herrmann C (2018) Integrated computational life cycle engineering—application to the case of electric vehicles. CIRP Ann 67(1):25–28

Chwang CP, Lee SN, Yeh JT, Chen CY, Chao DY (2002) Water-vapor-permeable polyurethane resin. J Appl Polym Sci 86(8):2002–2010

Ciroth A, Muller S, Weidema B, Lesage P (2016) Empirically based uncertainty factors for the pedigree matrix in ecoinvent. Int J Life Cycle Assess 21(9):1338–1348. https://doi.org/10.1007/s11367-013-0670-5

European-Commission (2011) International reference life cycle data system (ILCD) handbook—recommendations for life cycle impact assessment in the European context, 1st edn

Frischknecht R, Jungbluth N et al (2007) Ecoinvent 2 overview and methodology. Ecoinvent Report No1, 1(1):1–77. Retrieved from http://www.ecoinvent.org/fileadmin/documents/en/01_OverviewAndMethodology.pdf

Hallstedt SI, Isaksson O (2017) Material criticality assessment in early phases of sustainable product development. J Clean Prod 161:40–52. https://doi.org/10.1016/j.jclepro.2017.05.085

Hauschild MZ, Rosenbaum RK, Olsen SI (2018) Life cycle assessment—theory and practice. Life cycle assessment—theory and practice (1st ed). Springer. Copenhagen/Montpellier. https://doi.org/10.1007/978-3-319-56475-3

ISO (2006a) Environmental management—life cycle assessment—principles and framework (ISO 14040). ISO, the International Organization for Standardization, Geneva

Kappenthuler S, Seeger S (2019) From resources to research—a framework for identification and prioritization of materials research for sustainable construction. Mater Today Sustain 100009. https://doi.org/10.1016/j.mtsust.2019.100009

Notarnicola B, Puig R, Raggi A, Fullana P, Tassielli G, De Camillis C, Rius A (2011) Life cycle assessment of Italian and Spanish bovine leather production systems. Afinidad 68(553):167–180

Piccinno F, Hischier R, Seeger S, Som C (2016) From laboratory to industrial scale: a scale-up framework for chemical processes in life cycle assessment studies. J Clean Prod 135:1085–1097. https://doi.org/10.1016/j.jclepro.2016.06.164

Qiu LM, Sun LF, Liu XJ, Zhang SY (2013) Material selection combined with optimal structural design for mechanical parts. J Zhejiang Univ Sci A 14(6):383–392. https://doi.org/10.1631/jzus.A1300004

Ribeiro I, Peças P, Silva A, Henriques E (2008) Life cycle engineering methodology applied to material selection, a fender case study. J Clean Prod 16(17):1887–1899. https://doi.org/10.1016/j.jclepro.2008.01.002

Shibasaki M, Fischer M, Barthel L (2007) Effects on life cycle assessment—acale up of processes. Advances in life cycle engineering for sustainable manufacturing businesses—proceedings of the 14th CIRP conference on life cycle engineering, pp 377–381

Sun M, Rydh CJ, Kaebernick H (2004) Material grouping for simplified product life cycle assessment. J Sustain Product Design 3(1/2):45–58. https://doi.org/10.1023/b:jspd.0000035558.27697.02

Chapter 8
Analysis of Life Cycle Datasets for the Material Gold

Benjamin Fritz and Mario Schmidt

Abstract The representation of gold-producing processes in common life cycle assessment (LCA) databases is insufficient. The biggest problems identified are the missing data for recycling of high-value scraps and for ASM and the estimations in industrial mining. The life cycle inventories (LCI) for the latter are based on corporate reports. The data available from the company figures are always incomplete and must therefore be scaled between the different mines. This process was defined in this work as Intersystemic-Data-Scaling (IDS). An analogy is presumed here between mines, although literature shows that there are differences in mines like ore types that affect the extraction processes and thus the LCI. In the present study all the assumptions and IDS were visualized in a world map. It was found that except for energy demand and production volumes there is no flow without IDS. Finally, the actual shares of the different gold routes in the world market were estimated using literature research. When compared to the market shares used in common life cycle databases it can be seen that there are big data gaps emphasizing the importance of further data collection for the life cycle datasets for the material gold.

Keywords Life cycle inventories · Gold mining · Mineral extraction · Data gaps · Market datasets

8.1 Introduction

Gold has always been one of the most sought-after precious metals. However, the conditions in which the valuable mineral is mined are often kept quiet. Besides its various applications in the investment, jewelry and industrial sector gold also has

B. Fritz (✉) · M. Schmidt
Institute for Industrial Ecology (INEC), Pforzheim University, Tiefenbronner Straße 65, 75175 Pforzheim, Germany
e-mail: benjamin.fritz@hs-pforzheim.de

M. Schmidt
Faculty of Sustainability, University Lüneburg, Universitätsallee 1, 21335 Lüneburg, Germany

S. Albrecht et al. (eds.), *Progress in Life Cycle Assessment 2019*, Sustainable Production, Life Cycle Engineering and Management, https://doi.org/10.1007/978-3-030-50519-6_8

a bad reputation since it leads to negative impacts that are of great importance—resource depletion, the extensive use of chemicals, toxic emissions, high energy consumption and social concerns, just to name a few.

The gold stock already mined is estimated at approximately 190,400 metric tons or a cube with an edge length of 21 meters in 2018, of which 2/3 were mined after 1950. Of this figure, approximately 73,000 metric tons (40%) of gold are stored in bank safes and vaults World Gold Council (2019a). In 2018, the demand for gold was 1800 metric tons for investment and 2,600 metric tons for the manufacture of jewelry and industrial goods. In addition, demand has been fairly constant for the last decade World Gold Council (2019b). If gold stocks from banks and vaults were used to meet the demand for gold for jewelry and industrial goods, the world could live 28 years without gold from mines. Furthermore one could include some values from the literature on the environmental impacts of gold mining in this thought experiment like global warming potential of gold from mines with approx. 20,000 kg CO_2-equivalent per kg gold (kg CO_2-eq/kg Au) World Gold Council (2018a) or Cyanide demand of approx. 140 kg cyanide/kg Au (Li 2013). The environmental impacts of this thought experiment would lead to a saving of incredible 1.5 Gt CO_2-eq or the potential risks of 11 Mt cyanide could be avoided.

With this simplified example, it can be shown how different the discussion on environmental issues in gold production is from other resources. Gold is mainly used for luxury, decoration, status and investment and mainly remains highly concentrated in the system, unlike other mining resources such as coal or metals, which are subject to high dissipation or even chemical transformation. In detail, the application of gold is shown in Fig. 8.1.

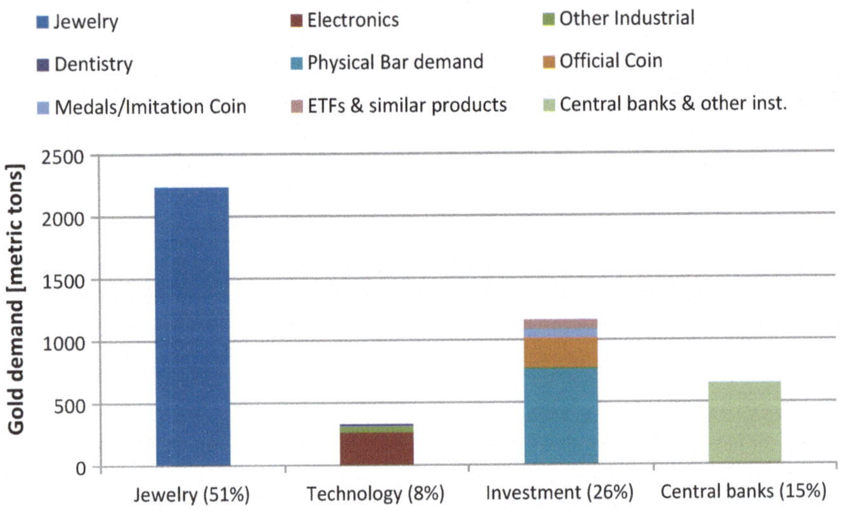

Fig. 8.1 World gold demand by different applications (World Gold Council 2019b)

The demand for these various applications is covered by the production of gold. There are different approaches to subdivide the different production routes. In the area of life cycle assessment (LCA), it makes sense to subdivide according to process technologies. Figure 8.2 shows the production of gold according to its main process routes.

The ten countries that mined the most gold in 2018 can be found in Fig. 8.3. In 2005, South Africa, Australia and the USA led the production statistics (CPM Group

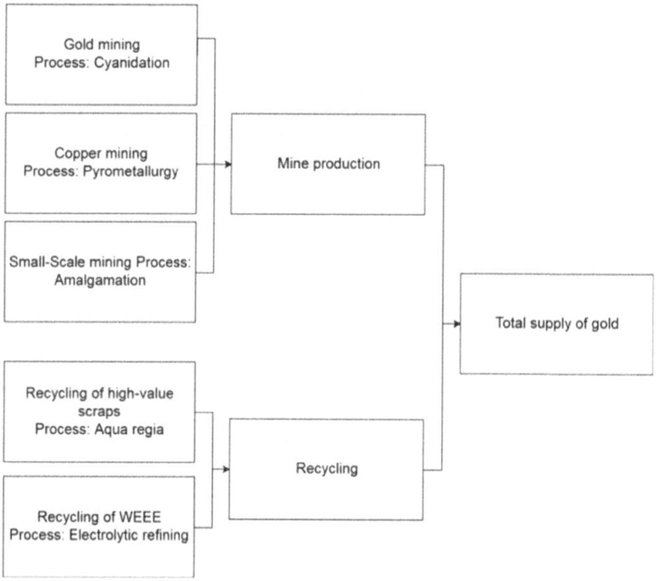

Fig. 8.2 Gold production by different process technologies

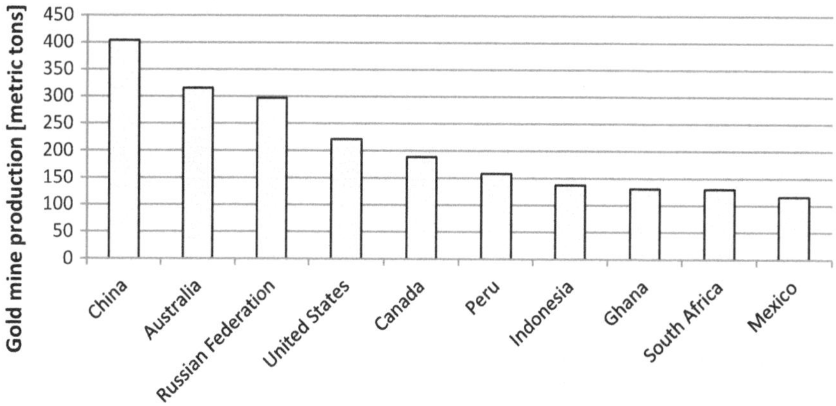

Fig. 8.3 Gold production in 2018 by different countries (World Gold Council 2019b)

2018). In 2018 China, Australia and Russia were the three countries with the largest production of gold from mining World Gold Council (2019b).

Gold mining is associated with major environmental impacts. This has a major impact on the LCA of electronic products. In a study by Ercan and Malmodin (2016) on the LCA of smartphones, gold, based on ecoinvent data, accounts for more than 40% of the total impact in five of twelve impact categories. An LCA study on notebooks by O'Connell and Stutz (2010) based on GaBi data came to similar conclusions. By far the largest contribution to the GWP is made by the mainboard, in which the gold pins of the RAM bars contribute around 40%. Even such small amounts of gold, which are used in laptops or smartphones, already have an enormous impact on the overall product. It is therefore important to use the most accurate datasets possible in LCA. This study will therefore analyze the current LCA datasets on the material gold and examine the sources and quality of their content.

8.2 Methodology

The most widely used LCA databases are GaBi databases from the German company thinkstep AG and ecoinvent from a consortium around the Swiss Federal Institute of Technology Zurich (ETH Zurich). Hence, for ecoinvent, the data transparency is considerably higher due to detailed reports which allows analyzes in greater depth, it was the main source for this research. Nevertheless, some interesting insights from the information that is publicly accessible as well as from private communication with providers could also be gained for the GaBi database. Since the introduction of the first gold dataset in ecoinvent in 2007, all assumptions and improvements from the official ecoinvent reports have been investigated in a systematic literature research. The first report, Classen et al. (2007), in which gold appears, was analyzed in detail, particularly with regard to the assumptions and estimates made.

All subsequent change reports were then examined for changes affecting the datasets for gold. The description of the datasets from the databases is divided into the different production routes of gold by process technology (see Fig. 8.2). Table 8.1 also shows a more detailed list of all the change reports taken into account for the literature review.

Finally, the share of production routes in the global market was reviewed in the databases and compared with the actual composition of the gold market.

Table 8.1 List of all ecoinvent (EI) documentation reports used for the literature research

Year	Author: Title
2007	M. Classen, H.-J. Althaus et al.: Life Cycle Inventories of Metals
2009	M. Classen, H.-J. Althaus et al.: life Cycle Inventories of Metals
2009	H.-J. Althaus, C. Bauer et al.: Documentation of changes implemented in EI Data v2.1
2010	H.-J. Althaus, C. Bauer et al.: Documentation of changes implemented in EI Data v2.1 and v2.2
2013	E. Moreno Ruiz, B.P. Weidema et al.: Documentation of changes implemented in EI Data 3.0
2014	E Moreno Ruiz, T. Lèvovà et al.: Documentation of changes implemented in EI Data 3.1
2015	E Moreno Ruiz, T. Lèvovà et al.: Documentation of changes implemented in EI Data 3.2
2016	E Moreno Ruiz, T. Lèvovà et al.: Documentation of changes implemented in EI Data 3.3
2017	E Moreno Ruiz, L. Valasina et al.: Documentation of changes implemented in EI Data v3.4
2018	E Moreno Ruiz, L. Valasina et al.:Documentation of changes implemented in EI Data v3.5

8.3 Analysis of Common LCA Datasets

8.3.1 Industrial Gold Mining with Cyanidation

The process in GaBi, which represents the globally averaged situation for primary gold, consists of data from South Africa, Ghana, Peru and Australia. These four are among the top ten gold producing countries, accounting for 20% of world production World Gold Council (2019b).

In ecoinvent, the data is more comprehensive and can be analyzed in much more depth. The data stems from public reports by the world's most important gold producers in 2005. In addition to the countries Peru, Ghana, South Africa and Australia, which are also included in GaBi, Canada, Chile, Papua New Guinea, Sweden, Tanzania and the USA were included in ecoinvent as well. In gold mining, it is common practice to also publish technology and environmental reports along with the financial ones. With a combination of these three, it is often possible to derive some data such as electricity and water consumption or the quantity of explosives per quantity of gold produced. Normally, not all relevant data is available for every mine. In ecoinvent this problem is solved by transferring existing data from one mine to missing data of another mine. The reference value is usually the quantity of gold or ore mined. This means that some of the values in the life cycle inventories (LCIs), such as chemicals or emissions are adjusted from another source than the original mine by scaling them according to production volumes. In other words, ecoinvent assumes an analogy between the different mines. In this work, this process will be referred to as Intersystemic-Data-Scaling (IDS).

It is obvious that not all mines are identical and consequently assumptions and simplifications have to be made in the field of LCA. However, for mines some differences have a greater impact on LCI-relevant data than others. For example, underground and surface mines differ in their energy requirements due to differences in ore content and additional energy requirements for ventilation of the underground mines (Koppelaar and Koppelaar 2016). Refractory and non-refractory ores differ greatly in the preparation of the ores necessary for cyanide leaching and thus in the results of the LCA (Norgate and Haque 2012). Moreover the LCA results are affected by the fact of whether and what by-products are produced in a mine. Besides differences in processes, the often discussed allocation problem plays an important role here (Santero and Hendry 2016). Last but not least, the level of technological development and environmental policy in a country also has an influence on the environmental impact of mines Roche et al. (2017).

A closer look at the data reveals that the most complete dataset is gold production [ZA] in South Africa. However, for five of the total of nine country-specific gold production processes, more than half of the data stem from non-original datasets and therefore have more than 50% IDS. As an example the dataset of gold production [CL] in Chile has 85% IDS and consists apart from the production volume of gold and the level of energy consumption only of data from other mines in South Africa, Peru, Papua New Guinea and Sweden. Does such a dataset actually still reflect the specific gold production of a country? It is a common procedure in science to make well-founded assumptions, analogies and estimates. Especially in the field of LCA, where the calculated results are often only correct in the order of magnitude, but not to the last decimal place, such an approach can be applied. However, the term "gold production [CL]" suggests to the user that this is a dataset on gold production in Chile and not one based largely on data from other countries. It would make more sense to speak of a generic dataset representing e.g. a specific technology.

Looking at the individual LCIs by material category, it can be seen that the most IDS are found in the material type "chemicals". For the materials zinc and activated carbon, all data points of the nine countries represented in the datasets were scaled on the basis of a doctoral thesis by Stewart (1999) in which mines in South Africa were investigated. The same applies to the disposal of tailings. Here, all data points are taken from the Sustainability Reports of two mines in Papua New Guinea that were formerly operated by Placer Dome Inc. But both of these mines used a rather uncommon practice because the tailings were disposed of in water. Especially the topics "tailings" and "chemicals" enjoy great media attention. The widely quoted study by earthworks (Septoff 2004) stated, for example, that a wedding ring produces approximately 20 tons of toxic waste. Exactly these figures should be used with caution and, if necessary, questioned.

The example of the mines in Papua New Guinea can also be used as an example to pick up again the question of analogy between different mines. The two mines in Papua New Guinea, a country that has often been criticized for its environmental policy in mining, are two open-pit gold-silver mines with relatively small production volumes and mainly refractory ores (Nelson 2016; Porgera Joint Venture and Mining Operation 2019). Based on this dataset, IDS is now applied on various other mines

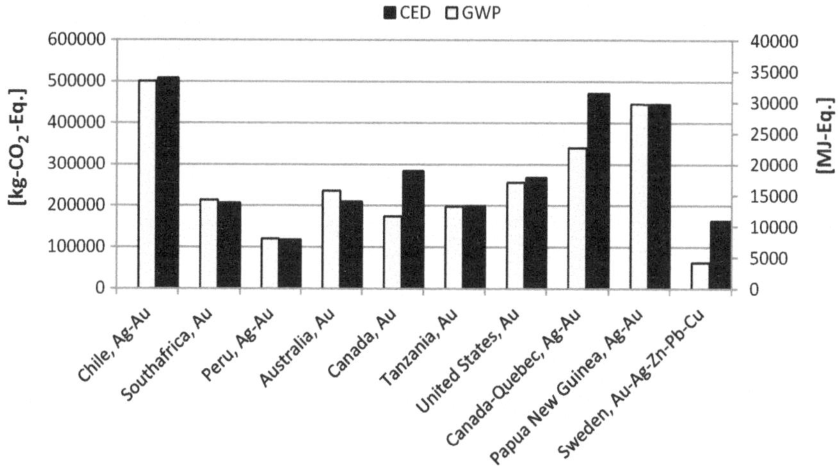

Fig. 8.4 Comparison of GWP and CED (per kg gold) for mining datasets of different countries

in different countries. When performing an environmental analysis to compare the Global Warming Potential (GWP) and the Cumulative Energy Demand (CED) for the datasets of all countries (see Fig. 8.4), the results are very different. Since GWP and CED in mining are mainly dependent on energy demands and these are the only values that were available for all datasets in the environmental reports without IDS, this picture is not particularly surprising. It supports much more the statement that there are large differences between the different mine datasets, which is why the analogy assumed by ecoinvent for many of the materials should be critically questioned.

All IDS are displayed on a world map in Fig. 8.5. This Figure underlines impressively the influence of the mines in South Africa and Papua New Guinea on the LCI's of the other country specific datasets. Furthermore, one can see that there is not a single dataset that does not include IDS. It is also apparent that no IDS for production volumes and energy demand for mining were performed since these values were available in all the environmental reports of the mines. Data on CO_2 emissions from explosions and the quantity of machines and buildings are missing for all mines and have therefore been scaled from literature values to all mines on the basis of the quantities of gold or ore produced. For refining, energy data were missing for six countries, which is why ecoinvent estimated an energy consumption of 1.63 kWh/kg Au according to literature values by Auerswald and Radcliffe (2005) and Renner et al. (2012).

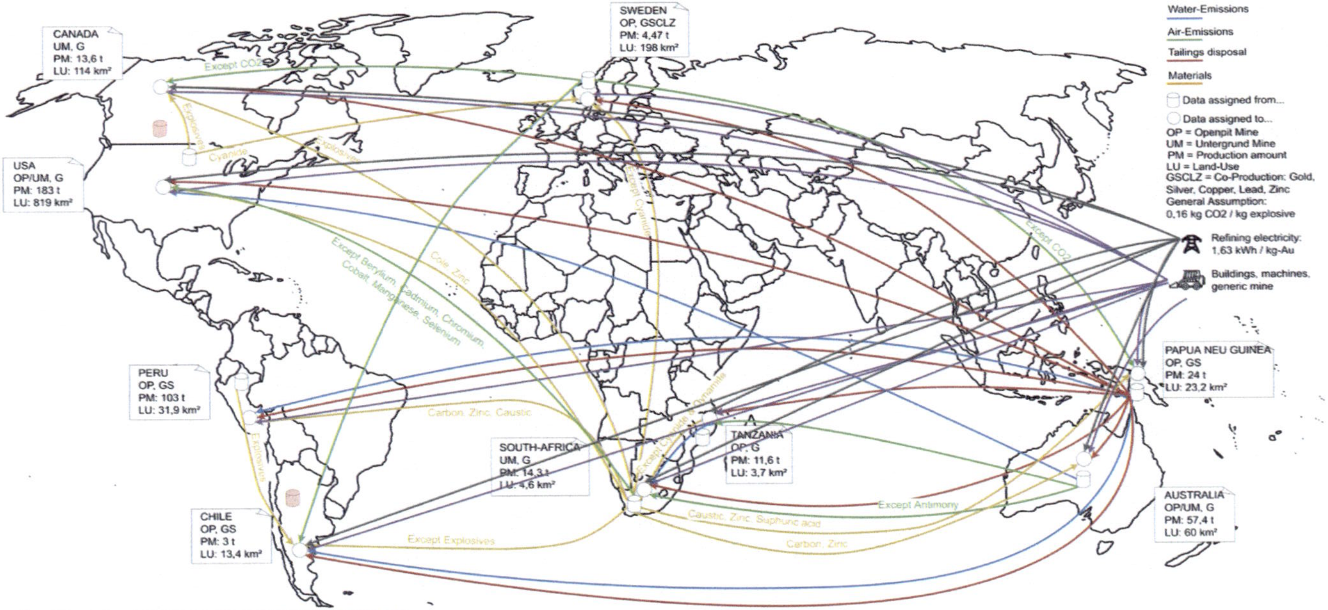

Essential productflows in the ecoinvent database for the mine production of gold.
Own figure based on Classen, Althaus, et. Al. (2009) LCI of Metals

Fig. 8.5 Intersystemic-data-scaling of ecoinvent dataset for gold mine production displayed on a world map

8.3.2 Generic Gold from World Market

Both GaBi and ecoinvent contain datasets that are supposed to represent gold from the world market. Market activities try to model a generic product by taking into account the globally averaged situation, focusing on the main technologies, the region specific characteristics and/or import statistics. Here, the routes shown in Fig. 8.2 and the production countries shown in Fig. 8.3 are included by world market share. In both databases the data for ASM are completely missing. Also for the route of high-value gold scrap recycling there are currently no data available. In GaBi the route appears in its market activity with a share of 27%, but it is assumed as a simplification that high-value gold scrap recycling is only a melting process of gold (private communication). Until ecoinvent v.3 this data gap was closed by assuming that Waste Electronic Equipment (WEEE) recycling has the exact same impact as high-value gold scrap recycling and has a market share of 30% (Classen et al. 2009). In ecoinvent v.3.5 high-value gold scrap recycling was omitted, so that the share is 1% from WEEE and 99% from mining (private communication). According to Adams (2016), a common process for high-value gold scrap recycling is hydrometallurgical treatment with aqua regia. In ecoinvent the above-mentioned process of gold from WEEE recycling is based on data from a plant in Sweden and contains besides the production volumes only IDS. All the LCI data of this dataset, like energy consumption or emissions is based on assumptions and literature values of furnaces and electrolysis plants. In the GaBi database there is a value for WEEE recycling but the process is not included in the documentation. The GaBi database has a process which represents the production of gold from copper ores with a market share of 20%. The data is based on a plant by Copper Refineries Ltd. in Townsville, Australia, which first electrometallurgical refines the copper and then extracts the gold from the anode slimes by chlorination. In ecoinvent this route is not shown separately, but one of the mines considered in the datasets for gold production from mining is a copper mine with a copper smelter (see Fig. 8.5). An overview of the quantities of the individual routes in the market activity of ecoinvent and GaBi can be found in Fig. 8.5.

In the following it shall be attempted to estimate the actual shares of gold production by different process technologies on the basis of literature and market statistics. An annual worldwide mine production of 3501 metric tons and a production from recycling of 1168 metric tons is assumed World Gold Council (2019b). Furthermore, a market report by World Gold Council (2018b) referring to the study by Alistair et al. (Hewitt et al. 2015) states, that recycling is divided into 90% from high-value gold scrap and 10% from electronic waste (WEEE). For ASM there are many different studies with different estimates. In the present work the estimation by Seccatore et al. (2014) was used because it is the only study identified that explains in detail and very transparent how the estimation was done. The study comes to a value between 380 and 450 metric tons. Since the gold price and the mining areas increased since 2014 it can be assumed that production most likely increased since 2014 and therefore the upper value of 450 metric tons was chosen (CPM Group 2018; Lobo et al. 2018). It is difficult to determine how many of these 450 metric tons from ASM is accounted

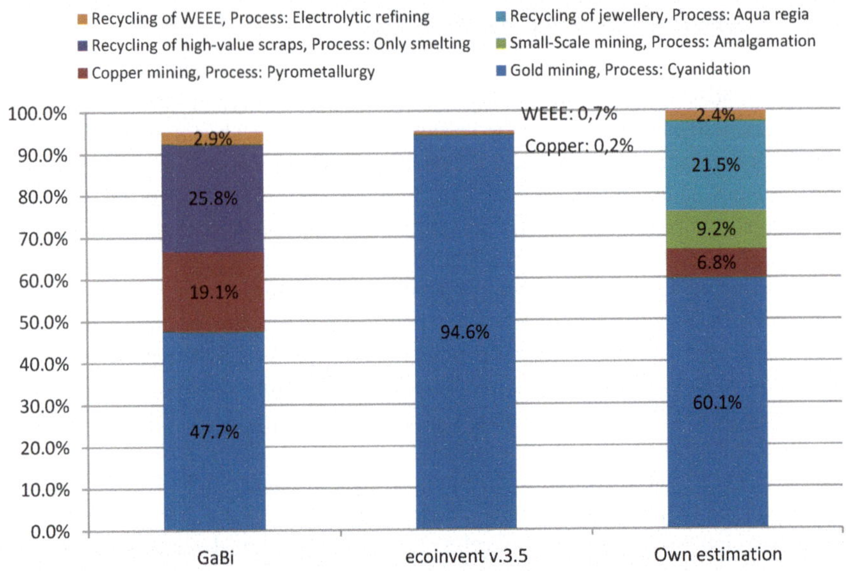

Fig. 8.6 Comparison of global shares of different gold production processes in different LCA databases with market and literature values. To illustrate the amount of gold from ASM traded outside the official world market the bars *GaBi* and *ecoinvent v.3.5* sum up to less than 100%

for in the official market statistics and how much is sold on the quiet. Based on personal discussions with experts and own observations from ASM gold mines in the Brazilian Amazon rainforest, it can be assumed that a large part of this gold is traded outside the official world market. For this study it is assumed that 50% of the ASM is sold on the official markets and therefore the total mine production of 3501 metric tons has to be increased by 225 metric tons.

According to Butterman and Amey (2005) the proportion of gold from copper concentrate is between 5 and 15%. Again the higher value of 15% was chosen as since 2005 the gold price and the Chinese production have increased. In Fig. 8.6 the results of the estimation on the actual shares of gold production by its different process technologies compared to their representation in LCA databases are shown. Note that the bars representing the database of GaBi and of ecoinvent v.3.5 are lower than 100% due to the amount of gold produced by ASM traded outside the official world market.

8.4 Conclusion

Gold has a very bad image from a social and environmental perspective. To quote an outlook on gold in Nature magazine in 2012 goldmining was called "unwanted neighbor" and "Gold not green" (Owens 2013). Recent reports from Brazil do not give a better picture of the social conditions of gold prospectors and of the decimation of the rainforest.

Looking at LCA of products, e.g. electronic products, gold has a very high environmental impact even in very small quantities. But gold comes from different production routes and countries, whose social and ecological problems are very different. The most important production routes for gold are: mining with cyanidation, pyrometallurgic refining of copper ores, small scale mining with amalgamation, recycling of high-value scraps with aqua regia and electrolytic refining of WEEE. In the context of this article, these different routes were studied in more detail and their inclusion in the most widely used LCA databases GaBi and ecoinvent were examined.

The dataset for gold mining with cyanidation is in the GaBi database based on datasets from four countries while in ecoinvent there are nine. These data records in ecoinvent originate exclusively from company reports and literature values. The majority of these company figures are no longer publicly available and cannot be verified. The data available from the company figures are always incomplete and must therefore always be scaled back and forth between the data of the LCIs. In this study, this is referred to as Intersystemic Data Scaling (IDS). All these assumptions and IDS for all nine country datasets on gold mining were graphically depicted on a world map (see Fig. 8.5). In this graphic presentation one can see how dominant the datasets of the mines in South Africa and Papua New Guinea are for chemicals and tailings disposal. Apart from the fact that many of the company figures can no longer be found today, there is another problem with the company figures. Classen et al. (2009) describe this problem as follows: "Data in this report are based mostly on environmental reports of large multinational companies. However, it must be assumed that these sources represent rather the best practices for gold mining". With ore grades ranging from five to 40 g of gold per ton of ore one can see two things. First, for even tiny amounts of gold huge pits or deep shafts have to be dug which needs a lot of energy. Second, different gold mining operations can have very different environmental impacts. Therefore inaccurate data has huge impacts on the results of different LCA studies containing the material gold.

For gold from copper ores GaBi has a specific dataset for the electrolytic treatment of anode slimes. In ecoinvent there is no specific dataset for this, but one of the mines in the dataset for gold production from mining is a copper mine which produces gold only as a by-product. This study showed that the gold from copper ores makes up for around 7% of the worlds gold production (see Fig. 8.6).

Gold from Artisanal and Small-Scale Mining (ASM) is not found in any of the LCA databases, although it plays a major global role and there are many studies in the field of environment and sustainability on this topic. In ASM, other and additional environmental impacts such as mercury emissions or sediment inputs into rivers are most probably a big problem and therefore further work needs to be done to integrate these process routes in LCA databases (UNEP 2013; Lobo et al. 2016).

In ecoinvent, the process that represents gold from WEEE recycling is based on the production volumes of a plant in Sweden. All energy consumption and emissions of furnaces and electrolysis plants are based on assumptions and literature values

(Classen et al. 2009). Especially this topic is getting a lot of attention right now by scientists and also the media although it is only responsible for about 10% of the amount of gold produced from recycling. The other 90% stems from recycling of jewelry, coins and other so called high-value gold scrap.

There is currently no LCA data record for high-value gold scrap recycling. GaBi assumes a simple melting process for this in its market activity and ecoinvent has omitted this route. The literature shows that a common process for this is hydrometallurgical treatment with aqua regia. Our results show that this route is responsible for more than 20% of the worlds gold production. The environmental impacts of this route on the other hand compared to WEEE and/or any mining activities are most probably significantly lower. Simply because the gold content of the input material is already much higher.

In most cases, the value chain of gold is not known, which is why LCA practitioners often use the generic LCA processes also referred to as market processes for their product LCA's. If one takes a look at the examples of product LCA's mentioned in the introduction where even the smallest amounts of gold have a significant effect on the examined products, it is important to analyze and understand these generic datasets. On the basis of market statistics and literature research, an attempt was made to estimate the actual market shares of the different gold routes as transparently as possible. When compared to the market shares used in GaBi and ecoinvent it can be seen that there are big data gaps emphasizing the importance of further data collection for the life cycle datasets for the material gold.

8.5 Outlook

To develop a full picture of the environmental impacts of the material gold, additional studies will be needed in the field of life cycle assessment in relation to gold. More primary data is needed on mine extraction in different countries and on possible recycling routes. There is still no data published on the recycling of high-value gold scraps, which accounts for at least 20% of the gold produced worldwide. The ecoinvent data on gold from WEEE recycling are based on incomplete data from only one recycling plant. New and reliable data must also be collected in the field of industrial mining, especially for the disposal of tailings and the use of chemicals. Both the ecoinvent and the GaBi dataset on gold from mining have one thing in common: the data are based almost exclusively on voluntary company reports and must therefore be viewed very critically. The frequent discussion on spatial differentiation in life cycle impact assessment (LCIA) to increase the environmental realism of LCIA is particularly relevant in the field of mining (Hauschild 2006). The Chinese market, for example, is missing in the databases even though it has been the world's largest gold producer for years now and according to Chen et al. (2018) is associated with almost double the GWP then commonly used values. The lack of countries such as Burkhina Faso or Brazil, which are among the top 20 most gold producing countries and where ASM is important, also shows how important it is to collect ASM data. So far, three LCA studies on ASM gold in Peru and the Philippines have been identified (Cenia et al. 2018; Kahhat et al. 2019; Valdivia and Ugaya 2011). In order to close this gap and include these data in common datasets, it is important to collect more data in different regions. Further research is also needed to determine the market share of ASM of gold as accurately as possible. This is also part of a more general problem of transparency in the value chains of gold that could be solved e.g. by using blockchain techniques.

References

Adams MD (ed) (2016) Gold ore processing. Project development and operations, Second edition. Elsevier, Amsterdam, Boston, Heidelberg, London

Auerswald DA, Radcliffe PH (2005) Process technology development at Rand Refinery. Miner Eng 18(8):748–753

Butterman WC, Amey EB (2005) Mineral commodity profiles—gold. US Geological Survey

Cenia MCB, Tamayao M-AM, Soriano VJ, Gotera KMC, Custodio BP (2018) Life cycle energy use and CO2 emissions of small-scale gold mining and refining processes in the Philippines. J Cleaner Prod 23(10):1928–1939

Chen W, Geng Y, Hong J, Dong H, Cui X, Sun M, Zhang Q (2018) Life cycle assessment of gold production in China. J Cleaner Prod 179:143–150

Classen M, Althaus H-J, Blaser S, Scharnhorst W, Tuchschmid M, Jungbluth N, Emmenegger MF (2007) Life cycle inventories of metals data v2.0. ecoinvent, Dübendorf

Classen M, Althaus H-J, Tuchschmid M, Jungbluth N (2009) Life cycle inventories of metals data v2.1. Part IX - Gold and Silver. ecoinvent, Dübendorf

CPM Group (2018) The CPM gold yearbook 2017, 1st edn. Wiley Trading Series. Wiley, J, New York, NY

Ercan M, Malmodin J (eds) (2016) Life cycle assessment of a smartphone. ICT for sustainability 2016, Amsterdam, the Netherlands, 30.08.2016–01.09.2016. Atlantis Press, Paris, France 2016

Hauschild M (2006) Spatial differentiation in life cycle impact assessment: a decade of method development to increase the environmental realism of LCIA. Int J LCA 11(S1):11–13

Hewitt A, Keel T, Tauber M, Le-Fiedler T (2015) The ups and downs of gold recycling: understanding market drivers and industry challenges. https://www.bcg.com/de-de/publications/2015/metals-mining-cost-efficiency-ups-and-downs-of-gold-recycling.aspx. Accessed 06 2020

Kahhat R, Parodi E, Larrea-Gallegos G, Mesta C, Vázquez-Rowe I (2019) Environmental impacts of the life cycle of alluvial gold mining in the Peruvian Amazon rainforest. Sci Total Environ (662):940–951

Koppelaar RHEM, Koppelaar H (2016) The ore grade and depth influence on copper energy inputs. Biophys Econ Resour Qual 1(2):9

Li C, Li H, Wang M (2013) Life cycle assessment of different gold extraction processes. In: Energy technology 2014 - carbon dioxide management and other technologies. Wiley, J, Hoboken

Lobo F, Costa M, Novo E, Telmer K (2016) Distribution of artisanal and small-scale gold mining in the Tapajós River Basin (Brazilian Amazon) over the past 40 years and relationship with water siltation. Remote Sens 8(7):579

Lobo F, Souza-Filho PWM, Novo EMLdM, Carlos FM, Barbosa CCF (2018) Mapping mining areas in the Brazilian Amazon Using MSI/Sentinel-2 Imagery (2017). Remote Sens 10(8):1178

Nelson H (2016) Black, white and gold. Goldmining in Papua New Guinea 1878–1930. ANU Press, Acton, Australia

Norgate T, Haque N (2012) Using life cycle assessment to evaluate some environmental impacts of gold production. J Cleaner Prod 29-30:53–63

O'Connell S, Stutz M (2010) Product carbon footprint (PCF) assessment of Dell laptop - Results and recommendations. In: Proceedings of the 2010 IEEE International Symposium on Sustainable Systems and Technology. IEEE, pp 1–6

Owens B (2013) Mining: extreme prospects. Nature 495(7440): S4–6

Porgera Joint Venture, Mining Operation (2019) http://www.porgerajv.com/Our-Operation/Mining-Operation

Renner H, Schlamp G, Hollmann D, Lüschow HM, Tews P, Rothaut J, Dermann K, Knödler A, Hecht C, Schlott M, Drieselmann R, Peter C, Schiele R (2012) Gold, gold alloys, and gold compounds. In: Elvers B (ed) Ullmann's encyclopedia of industrial chemistry, vol 84. Wiley-VCH Verlag GmbH & Co. KGaA, Weinheim, Germany, pp 94–139

Roche C, Thygesen K, Aker E (2017) Mine tailings storage: safety is no accident. A UNEP Rapid Response Assessment. United Nations Environment Programme; GRID-Ardenal, Nairobi, Arendal

Santero N, Hendry J (2016) Harmonization of LCA methodologies for the metal and mining industry. Int J LCA 21(11):1543–1553

Seccatore J, Veiga M, Origliasso C, Marin T, Tomi G de (2014) An estimation of the artisanal small-scale production of gold in the world. Science of the Total Environment 496:662–667

Septoff A (2004) How the 20 tons of mine waste per gold ring figure was calculated. https://earthworks.org/cms/assets/uploads/archive/files/publications/20TonsMemo_FINAL.pdf. Accessed Aug 2019

Stewart M (1999) Environmental life cycle considerations for design-related decision making in minerals processing (Thesis). University of Capetown, Capetown

UNEP (2013) Minamata convention on mercury, New York

Valdivia SM, Ugaya CML (2011) Life cycle inventories of gold artisanal and small-scale mining activities in Peru. J Ind Ecol 15(6):922–936

World Gold Council (2018a) Gold and climate change: an introduction. World Gold Council, United Kingdom

World Gold Council (2018b) Gold market primer—gold recycling. World Gold Council, United Kingdom

World Gold Council (2019a) How much gold has been mined? https://www.gold.org/about-gold/gold-supply/gold-mining/how-much-gold. Accessed 10 2019

World Gold Council (2019b) Gold mining production volumes. https://www.gold.org/goldhub/data/historical-mine-production. Accessed 10 2019

Chapter 9
Suggestions for the Technical Integration of Life Cycle Assessment Data Sets of ÖKOBAUDAT into Building Information Modeling and Industry Foundation Classes

Sebastian Theißen, Jannick Höper, Reinhard Wimmer, Anica Meins-Becker, and Michaela Lambertz

Abstract The Life Cycle Assessment (LCA) of buildings is an important evaluation method for the environmental quality of a building and its impact on climate and environment. However, the efforts required are very high and cost-intensive. Whole-building LCAs are therefore usually only used as theoretical evidence after completion of a certified building. The potential for full environmental optimization in early project phases is therefore not used in practice. New digital planning methods such as Building Information Modeling (BIM) offer the possibility of integrating and linking much of the information required for a whole-building LCA when creating a digital building model, which should be incorporated by all participants in the planning right from the start. Technical, organizational and contractual prerequisites must be (further) developed so that a whole-building LCA can be applied earlier and more easily with the aid of the BIM method. Within the framework of this work, primarily issues and their technical improvements of the official German LCA database ÖKOBAUDAT and German green building certification systems BNB/DGNB are presented. Based on the open BIM standard of the IFC data model, current fundamentals will be examined to adapt ÖKOBAUDAT aiming to simplify BIM integrated calculation of whole-building LCAs. In this context, concrete recommendations for action to extend the IFC data model and ÖKOBAUDAT will be discussed.

Keywords Whole-building life cycle assessment · Building information modeling (BIM) · Industry foundation classes (IFC)

S. Theißen (✉) · J. Höper · M. Lambertz
TH Köln (University of Applied Sciences), Cologne, Germany
e-mail: sebastian.theissen@th-koeln.de

R. Wimmer
TMM Group, Böblingen, Germany

A. Meins-Becker
University of Wuppertal, Wuppertal, Germany

S. Albrecht et al. (eds.), *Progress in Life Cycle Assessment 2019*,
Sustainable Production, Life Cycle Engineering and Management,
https://doi.org/10.1007/978-3-030-50519-6_9

9.1 Introduction

BIM is a holistic and collaborative working method that enables the digitization of the entire structure from "cradle to grave" and serves the most diverse perspectives of actors for planning, analysis, construction, operation and optimization. The depth of information in digital buildings grows through the integration of project-specific data.

Many software manufacturers have brought proprietary BIM systems onto the market, but due to a lack of interoperability they do not represent the universal solution for all user groups.

Since 1995, the organization buildingSMART has been developing the manufacturer-independent open source data format Industry Foundation Classes (IFC), standardized in DIN EN ISO 16739 (2018). It has the potential to establish itself as the standard compared to the many individual solutions. This interoperable data format offers interfaces beyond the numerous software boundaries and allows the exchange of data among all different disciplines on the basis of a digital BIM model

The IFC data format is structured in such a way that it allows dynamically expandable data sets. Based on this IFC extension scheme "Product Extension", information sources can be assigned to the objects or properties of the data model (IfcRelAssociates, IfcPropertySet), which can be linked to each other. So far, the extensions address material data (e.g., thermal conductivity) and economic aspects (time and cost planning). They are also suitable for mapping more complex relationships (e.g., sustainability).

Certification systems, such as the German assessment system for sustainable buildings (BNB) (Bundesministerium für Umwelt, Naturschutz, Bau und Reaktorsicherheit (BMUB) 2019) or the system of German Sustainable Building Council (DGNB) (2018), play a particularly important role within the environmental assessment of buildings today. LCA serves as the basic calculation method for environmental analysis. The environmental impacts of buildings are determined on the basis of verified LCA data by the Federal Ministry of the Interior, Building and Home Affairs (BMI, ÖKOBAUDAT) (Figl et al. 2017).

Currently, carrying out a whole-building Life cycle Assessment (LCA) is quite complex and time-consuming. Energy and material flows must be taken manually from 2D drawings and building descriptions. In particular, the technical building services equipment in the context of LCA is usually only simplified due to the high expenditure of a detailed consideration. Thus, the full optimization potential of an LCA as an early and iterative planning instrument cannot be used.

BIM offers a high potential to make the preparation of whole-building LCA much more efficient. By having the information required for the calculation earlier, and making it more structured and more easily accessible, an almost fully automated whole-building LCA would be possible.

To this purpose, technical, organizational and contractual (new) principles must be further developed. On the technical side, this means completely integrating and linking LCA data and further calculations with BIM model-specific parameters. The evaluation including its visualization and communication of LCA results must also be placed under the technical requirements. The organizational requirements include the definition of all interfaces, roles and responsibilities as well as the recording of all tasks for the whole-building LCA process over the life cycle. The purpose is to describe or subsequently standardize all information delivery and transfer times between the process participants with the definition of the respective necessary data quantities and detail levels with the aid of an Information Delivery Manual (IDM) in accordance with DIN EN ISO 29481 (2018). From a contractual point of view, the knowledge gained must be able to be integrated as a basis in Exchange Information Requirements (EIR) and BIM Execution Plan (BEP). This provides assistance in the form of modeling guidelines, among other things.

Within the scope of this work, primarily solutions for the technical prerequisites for the integration of life cycle assessment data sets are presented.

9.2 Basic of Whole-Building LCA

Here following, the norms and standards that form the basis for calculating the environmental impacts of building constructions and technical building equipment are described. In general, a life cycle assessment (LCA) describes a method with the purpose to provide information on how a product or process affects the environment throughout its life cycle. This makes possible to measure and compare environmental impacts in the various phases of a life cycle.

9.2.1 ISO 14000 Series, DIN EN 15978 and DIN EN 15804

Within the ISO 14000 series, there are two standards that standardize the general life cycle assessment methodology. ISO 14040 describes the most important principles and structure for carrying out an LCA (2009). ISO 14044 defines details and recommendations on how the LCA assessment process should be carried out (2006).

DIN EN 15978:2012 (2012) and DIN EN 15804:2014 (2014) are part of a series of standards that define the assessment of the environmental performance of buildings over their life cycle. All these standards are responsibility of the Technical Committee CEN/TC 350 Sustainability of Construction Work. While DIN EN 15978:2012 forms the basis for the whole-building LCA, DIN EN 15804:2014 defines product category rules for the environmental product declarations of construction products.

Also, DIN EN 15804:2014 standardizes which types of environmental impacts must be declared in which phases of a life cycle. A total of 24 different environmental indicators are defined, which are used in the life cycle phases: The environmental indicators are provided in the manufacturing phase (A1–A3), construction phase (A4–5), use phase (B1–7) and disposal phase (C1–4). The phases are supplemented by a module D, which balances information on reuse, recovery or recycling outside the life cycle and system boundaries (from cradle to grave). Currently only the declaration of modules A1–A3 is mandatory for the manufacturers of the products or respectively the producers of an EPD. The declaration of all other environmental impacts in the remaining life cycle phases is the voluntary choice of the manufacturer. Thus, for many data sets, for example, only environmental impacts from the manufacturing phase are available.

However, with the revision of DIN EN 15804:2018, the mandatory declaration of modules A1–A3, C1–C4 and D will soon also be introduced.

9.2.2 German LCA Database ÖKOBAUDAT and Data for Service Life

Life cycle assessment data sets are a central component of every life cycle assessment. They contain various types of information about construction products or processes that are important for calculating an LCA. In addition to environmental impacts, through environmental indicators, the information in the data sets usually also includes a description of the system boundaries, the declared unit and the area of application.

In Germany, freely accessible life cycle assessment data sets are available via the databases ÖKOBAUDAT and IBU.data. The ÖKOBAUDAT platform is made available as a standardized database for environmental assessments of buildings and is required as a data basis in the BNB and DGNB systems. Since September 2013, ÖKOBAUDAT has been the first life cycle assessment database to fully comply with the DIN EN 15804 standard (2014). It is sued nationally and internationally (Brockmann 2019), e.g. Denmark. A total of around 1200 building materials, construction and transport processes are currently described with regard to their environmental effects. Depending on the origin of the data sets, the following four data set types are distinguished:

- Generic Data: set collected from different sources with malus surcharge
- Representative: Average data set of selected manufacturers
- Average: Average data record of a manufacturer group
- Specific: data record of a manufacturer.

The ÖKOBAUDAT database consists to a large extent of generic data records. In Germany, the IBU (Institut für Bauen und Umwelt - Institute for Building and Environment) is making intensive efforts to provide more product-specific data sets for construction products in the form of standardized EPDs. This has been done since 2017 through the digital database IBU.data. One third of the 1200 LCA data sets of the current ÖKOBAUDAT are EPDs.

For the service lives of the building construction, BNB recommendations are used. The service life for technical building equipment is while taken from VDI standard 2067 (2012).

9.2.3 Performing a Whole-Building LCA According to BNB/DGNB

The German sustainability certification systems BNB/DGNB require the calculation of a whole-building LCA and define further calculation requirements based on DIN EN 15978:2012 and DIN EN 15804:2014. According to the DGNB requirements, the life cycle modules A1–A3 (production), B4 (refurbishment), B6 (energy demand during operational phase) and C3–C4 (waste) as well as module D (credits) must be calculated. ÖKOBAUDAT version 2016-I or newer is specified as the data basis (Bundesministerium des Innern, für Bau und Heimat (BMI) 2020). The technical building equipment can be integrated with two different calculation methods. The simplified procedure (VeV) and the complete procedure (VoV). While the VoV basically specifies a complete inclusion of all components of building constructions and technical building equipment, the VeV allows a limitation to eight essential component groups of both (Pohl 2014). The BNB system requires a very similar calculation of the whole-building LCA, without including module D into the calculation.

Within the scope of a BNB/DGNB certification, the environmental impacts listed in Table 9.1 are considered in accordance with DIN EN 15804:2014.

Table 9.1 Environmental Impact categories according to BNB/DGNB, based on DIN EN 15804:2014

Abbreviation	Indicator	Unit
GWP	Global warming potential	kg CO_2-Eqv
ODP	Ozone layer depletion potential	kg R11-Eqv
POCP	Photochemical ozone creation potential	kg Ethene-Eqv
AP	Acidification potential	kg SO_2-Eqv
EP	Eutrophication potential	kg PO_4-Eqv
PEges	PEges, sum of total use of non-renewable primary energy resources and total use of renewable primary energy resources (PEnrt + PErt = PEges)	MJ
PEnrt	PEnrt, total use of non-renewable primary energy resources	MJ
PErt	Total use of renewable primary energy resources	MJ

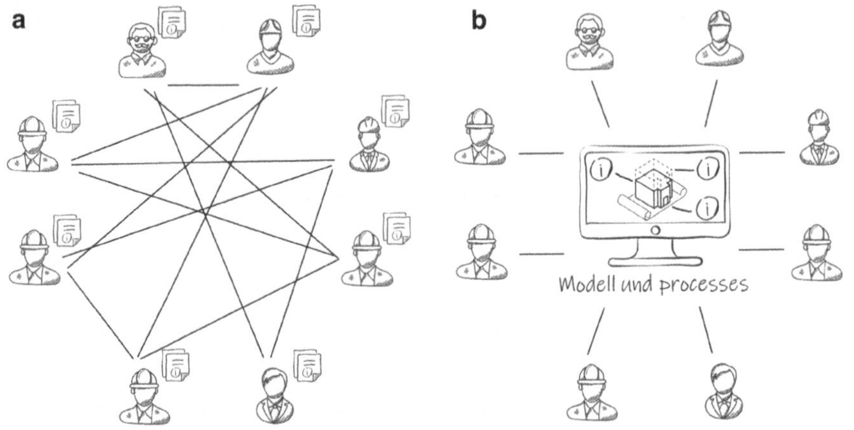

Fig. 9.1 **a** Usual information exchange within a design process. **b** BIM planning method enabling consistent information and data exchange

9.2.4 Building Information Modeling—Method

The BIM method is a digital integral working method in which continuous data enrichment takes place throughout the entire life cycle by integrating and linking all relevant building data in virtual data models. In this context, the individual phases are described, starting with the conception, planning and implementation up to the use and dismantling of a structure. The networking of the individual phases and the associated increase in the amount of information offer new opportunities for a consistent data exchange and standardized processes, as shown in Fig. 9.1b, compared to current information exchanges within a design process between the different stakeholders, visualized in Fig. 9.1a. The BIM method requires a technological architecture that guarantees the success of the project in a target-oriented manner.

Basically, they are called BIM modeling tools, BIM platforms or authoring tools. These extensive software tools enable the initial modeling of the 3D (BIM) model, enriched with data coming from many processes of the different stakeholders. In addition, so-called information management tools are used, which allow an extended processing of the data in the model to generate the relevant information.

9.2.5 Industry Foundation Classes

The basic data model in the context of open BIM applications is the Industry Foundation Classes (IFC) data model, standardized in DIN EN ISO 16739 (2017). IFC represents a manufacturer-independent, comprehensive standard for the description and exchange of digital building models. This data format was created in an effort to link different CAD tools beyond their proprietary boundaries and thus enable interoperability.

The IFC is a hierarchically object-oriented data format and that strictly distinguishes between semantic description and geometric representation. The classes in the IFC are uniquely technically defined and are formally mapped using object-related relations to other classes in addition to the inheritance relations. These definitions are generally concretized in the IFC data model by so-called entities, functions, rules, attributes and relations. In addition, there are the functions of the quantity sets and the property set (PSet), whereby dynamically extendable property sets can be defined modularly. Some of these property sets are already listed in the schema documentation that defines the data model. In this way, property records that go beyond the schema definition can be assigned to the individual classes independently of the specification, allowing an individualized semantic enhancement function. PSets are an important tool in the definition of object-inherent information. Data types and value ranges form the characteristics of the classes and can be further specified by enumerations and relations to other objects. For example, in the IFC model, the materials of the walls (IfcWall) are assigned to the individual layers (IfcMaterialLayerSet). The individual layers reflect the layer thickness and, in addition, reference is made to the definition of the actual materials using PSets from IfcMaterial. The Pset_EnvironmentalImpactIndicators (indicators for environmental impact) (buildingSMART International Limited 2019) and Pset_EnvironmentalImpactValues (values for environmental impact) (buildingSMART International Limited 2019) are particularly suitable as initial examination frames for the adaptation of ÖKOBAUDAT.

9.3 Current Shortcomings of IFC to Represent Environmental Impact Indicators/Values According to ÖKOBAUDAT and BNB/DGNB System

Based on the status quo of the IFC 4 data format (Addendum 2), the extent to which the structure can already be used to integrate life cycle assessment data from ÖKOBAUDAT will be examined. A distinction is made between the extent to which information for the German sustainability certification systems BNB/DGNB and for an ÖKOBAUDAT data set can be integrated.

The available IFC PSets "PSet_EnvironmentalImpactIndicators" and "PSet_EnvironmentalImpactValues" of the IFC 4 data format consist of the properties shown in Figs. 9.2 and 9.3.

The two PSets are primarily suitable for covering different LCA impact categories and their units. Information on a life cycle phase and an expected service life can also be included. Although this can already cover important life cycle assessment information for a BNB/DGNB life cycle assessment, there are many other important characteristics for the complete mapping of a life cycle assessment which are missing.

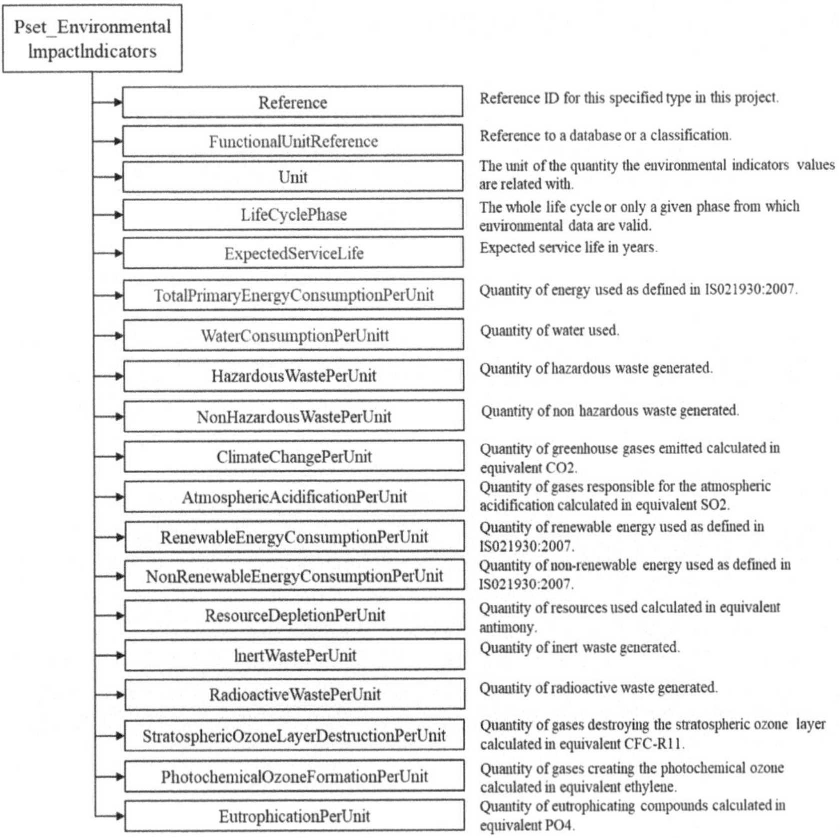

Fig. 9.2 IFC PSet "PSet_EnvironmentalImpactIndicators"

If the IFC PSets "PSet_EnvironmentalImpactIndicators" and "PSet_EnvironmentalImpactValues" are considered in terms of the complete representation of all ÖKOBAUDAT LCA data set information, the investigation shows that only a small proportion of the large amount of information can be integrated.

The integration of only one life cycle phase or life cycle module, which can only be assigned once by the LifeCyclePhase property, is particularly problematic. This means that it is currently only possible to integrate the environmental impacts of a certain phase (e.g., production A1–A3) or of a module (e.g., raw material extraction A1) through the property LifeCyclePhase. Assignments of several phases, modules, as required by the cradle to grave consideration according to BNB/DGNB (A1–A3, B2, B4, B6, C3–C4 (D)), are not possible. As consequence, the complete life cycle cannot be represented in the current IFC 4 data format.

Another aspect that restricts integration is the small number and types of LCA impact categories available in the IFC 4 format. In relation to the environmental

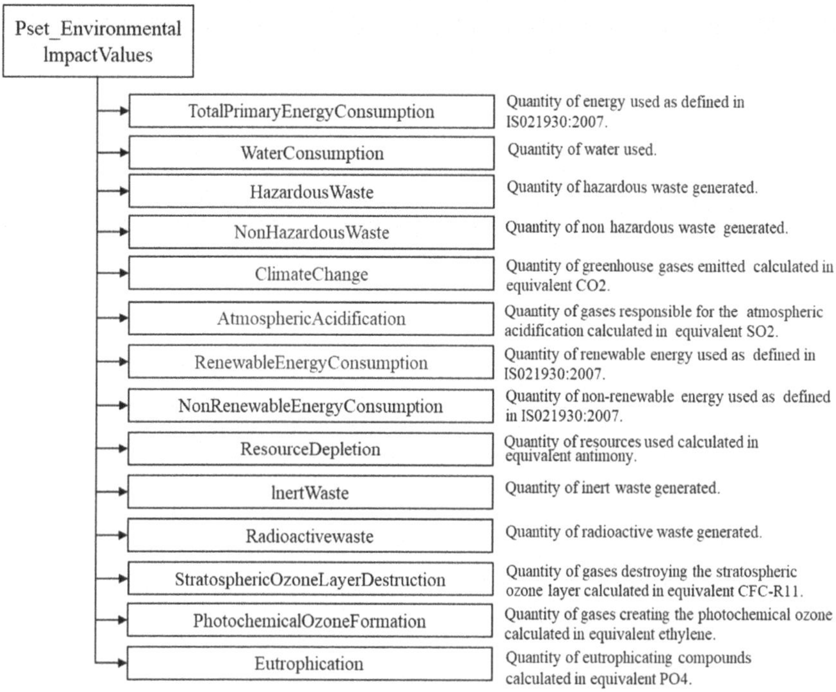

Fig. 9.3 IFC PSet "PSet_EnvironmentalImpact Values"

indicators required by BNB/DGNB, all of them can be mapped in IFC 4 format. As required by the BNB/DGNB, the units can also be equally integrated. In relation to LCA data sets of ÖKOBAUDAT, however, not all environmental indicators according to DIN EN 15804: 2014 or DIN EN 15804: 2018 (2018) are integrated.

Only 12 of the 24 required environmental indicators according to DIN EN 15804:2014 can be mapped in the IFC 4 data model so far. The PSet_EnvironmentalImpactIndicator/InertWastePerUnit cannot be assigned to any of the environmental impact categories or environmental indicators. On the other hand, hardly any other of the many information types of an ÖKOBAUDAT data set can be integrated into the BIM model with the PSets_EnvironmentalImpactIndicators/Values.

For example, the following parameters such as "Name", "Reference year", "Use advice for data set", "Technical purpose of product or process", "Reference flows", or "Technology description including background system" cannot be mapped via the two PSets.

9.4 Recommendations for Action and Solution Approaches to Integrate LCA Data Sets

Taking into account the current status quo of the IFC 4 data format for the integration of LCA data, recommendations for action and solution approaches are discussed with regard to their meaningful and realistic implementation to simplify a BIM-integrated automation of whole-building LCA. Therefore findings based on the research project are taken into account (Lambertz et al. 2020).

9.4.1 Communication via API Interface

A connection option, which offers direct communication with the API functions of ÖKOBAUDAT, enables synchronous access to the data of ÖKOBAUDAT. For communication with the API interface of ÖKOBAUDAT the software soda4LCA can be installed and used (Figl et al. 2017). Thus, the material data could be available to the user according to type and easily be integrated into the model by a link (e.g., UUID) in order to carry out an LCA on the basis of this.

The IFC 4 data format provides the concept "Classification Reference" for the reference of existing classifications. This concept ensures the functionality of the external reference of classifications into the BIM model. The connection is given by means of the IfcURIReference entity (buildingSMART International Limited 2019). As an example, the material (concrete of compressive strength class C50/60) can be integrated using the following reference: "https://www.oekobaudat.de/OEKOBAU.DAT/datasetdetail/process.xhtml?uuid=9fee07b9-a623-4e02-afc1-59637b0e1b7d."

This link communicates with a REST-API (Representational State Transfer—Application Programming Interface) and allows the communication of different systems. The documentation is done without data exchange and only requests are made that can be answered with data and information, which in turn can be processed. The responsibility for topicality and correctness remains guaranteed by the data base operator. Other relevant research projects also point out that a clear material and product classification could be stored as an external reference within BIM by means of a link to ÖKOBAUDAT or IBU.data as well as manufacturer catalogues in order to enable a clear allocation of a building material or product to its environmental profile (Gantner et al. 2018).

9.4.2 Integration of Service Life Within ÖKOBAUDAT

With the help of the service lives and their resulting replacement cycles, the consideration of the life cycle module B4 (replacement) is carried out alternatively according to the BNB/DGNB assessment. So far, the ÖKOBAUDAT data records do not contain any links to the service lives of the BNB system or the VDI 2067 standard. Only in a few cases is there information on the service lives available via the description of the background systems (e.g., 8.2.01 Fan central WRG 10,000 m^3/h, 20 years). It should therefore be discussed whether and how the integration of service lives required from BNB and VDI 2067 should take place.

Although the IFC 4 data format provides for two properties for the service lives (PSet_ServiceLife/ServiceLifeDuration or PSet_EnvironmentalImpactIndicators/ExpectedServiceLife), the implementation of the service lives from the two isolated sources for service lives presents itself as an additional problem. Since the service lives from the BNB and VDI 2067 standard are only available as PDF or Microsoft Excel files, which cannot be accessed via an interface, manual integration or assignment would be necessary in order to implement a BNB/DGNB compliant whole-building LCA within a BIM environment.

In addition, the type of modeling should be also considered. By using a layered modeling (e.g., the function "Create part elements" in AutoDesk Revit), a wall structure is created as a geometric body. At this regard, there are issues related, f.i., to the assignment of service life: IFC model allows indeed the export of only one value to be allocated into the geometry body by using a property function (in this instance "ExpectedServiceLife").

However, the several layers, which constitute a construction, may have different service lives. As a consequence, this type of modeling is very likely to present incorrect replacement cycles and thus incorrect consideration of replacement (B4-modul according to EN15084). A layer-by-layer modeling may solve this problem: individual layers of a wall structure are defined with separate assignments of e.g. layers service lives.

9.4.3 Unification of Quantitative Reference and Mandatory Field Gross Density

ÖKOBAUDAT currently offers a reference value and a reference unit for all data records. The reference value is usually related to 1 or less often to 1000. The reference value or declared unit is given in kg, m^3, m^2, m, number of pieces, kgkm, kWh, MJ, year depending on the building product, component or process. ÖKOBAUDAT also offers a gross density and a conversion factor for many data sets. Project-specific quantity, mass and volume data can therefore be converted when linking an ÖKOBAUDAT data set, depending on the existence and type of project-specific size, so that a correct linking of the environmental effects of the data set with the project-specific values is possible.

Modeling programs such as AutoDesk Revit automatically calculate the area and volume. The mass can also be calculated or exported. For this purpose, however, gross densities are used, which are stored internally or can originate from non-standard databases such as BIMobject. These can deviate in many cases from the ÖKOBAUDAT gross densities. It is therefore problematic if the mass of a wall from the BIM model (calculated with deviating gross density) is linked with an ÖKOBAUDAT data set that also contains the mass (kg) as reference value. Here it must either be ensured that the gross densities are identical in both cases or that the gross density of the ÖKOBAUDAT data set is used for all calculations. On the one hand, the integration of a new IFC 4 property would be necessary in order to integrate the gross densities of ÖKOBAUDAT and, if necessary, to make them comparable with the internal gross densities of the project within the BIM environment. On the other hand, this requires that material related LCA data sets of ÖKOBAUDAT provide information of gross densities. According to the current status this issue would require many adjustments in ÖKOBAUDAT.

A further problem under this aspect is the weight per unit area for the specification of ÖKOBAUDAT data records and the reference value m^2. For example, the data set for insulation (2.3.01 XPS Jackoboard) with a declared unit of 1 m^2 refers to a layer thickness of 18.5 mm, while a functionally equivalent data set (2.3.01 XPS alternate) with the same declared unit refers to a thickness of 100 mm and thus has a different condition. Such information can currently be found in various places in an ÖKOBAUDAT dataset and requires the LCA performer's specialist knowledge and efforts to view this information in the detailed information of the dataset.

In order to remedy such shortcomings, switching to two unified declared units could prove to be useful, as the modeling programs automatically calculate and specify this size at any time. In this case, the volume and the number of units could theoretically be used. Each ÖKOBAUDAT data set would be available in the unit m^3, whereby the consideration of processes, e.g., transport (kgkm) and use (kWh/MJ), is not taken into account here, since these are rarely used for the consideration of environmental impacts for the life cycle modules A1–A3, C1–C4 and D.

If ÖKOBAUDAT has an overall reference value of m^3, adjustments or conversions within the BIM environment are superfluous. In addition, information on layer thickness and weight per unit area in the ÖKOBAUDAT data records would no be longer required. This would also support a clearly low error potential in the consideration of the environmental impacts for the application case of an BIM integrated automation of whole-building LCA.

9.4.4 Supplementing Non-existent Life Cycle Modules (End of Life Processes)

In addition to the elementary adaptation of the IFC 4 data format for the integration of all life cycle phases or modules, there is also a need for further adaptation in terms of the integration of missing life cycle modules by the ÖKOBAUDAT data sets. Since, only the cradle to gate scenario (A1–A3) has been so far considered according to DIN

EN 15804:2014, many ÖKOBAUDAT data sets mainly lack end of life processes in order to implement a BNB/DGNB compliant whole-building LCA.

For example, the mineral wool process data set (2.1.01 Facade insulation) does not contain an end-of-life process. Therefore, an end of life process, a separate LCA data sets in ÖKOBAUDAT, must also be integrated additionally in the BIM Model. To solve this issue, an adaptation of ÖKOBAUDAT could be the extension of the existing generic data sets with end-of-life processes, which so far only cover the cradle to gate scenario (A1–A3). This extension would solve the problem with the correct linking of the life cycle modules in advance by ÖKOBAUDAT and would no longer pose a programming difficulty within the BIM applications. The eLCA tool from the Federal Institute for Research on Building, Urban Affairs and Spatial Development (BBSR) already implements this method and can therefore serve as a model (Rössig 2014). For example, the life cycle module C4—"disposal" process is automatically added for the facade insulation dataset mentioned above.

Therefore, before the draft of DIN EN 15804:2018 comes into force, which requires the mandatory consideration of A, C and D, an addition of life cycle modules in previous ÖKOBAUDAT data sets is considered as a sensible need for adaptation.

In a similar way, the draft of DIN EN 15804:2018 also results in adjustments with regard to environmental indicators. A breakdown of, e.g., GWP and EP will lead to new or more environmental indicators with partly new units. These adjustments must also be taken into account when adapting and integrating ÖKOBAUDAT.

9.4.5 Inclusion of EPDs and New Data Sets

The inclusion of new data records is also an adaptation of ÖKOBAUDAT. EPDs such as "Armaflex" or "CLIMAFLEX" pipe insulation could currently be included in the technical building equipment area in order to provide a broader or more product-specific data basis. In addition, it is proposed to integrate new data sets that have been missing so far or which did not appear relevant in the course of the two calculation procedures "simplified procedure" (VeV) and "complete procedure" (VoV) of BNB/DGNB. Since in practice almost exclusively the VeV method has been applied, there was no need to record distribution and transfer components of the technical building equipment in more detail in a whole-building LCA. Even building automation or electrical cables are generally not included in the life cycle assessment (Lambertz et al. 2019).

In particular, ventilation systems, such as ducts, outlets and molded parts, as well as components from the monitoring and control technology of building automation systems, should not be underestimated in terms of their environmental impact due to their material composition (plastics, metals) and short replacement cycles. When implementing an automated whole-building LCA, the distinction between VeV and VoV is omitted, since BIM models automatically provide all the necessary information on quantities, masses, quantities, etc. and only need to be linked to the appropriate LCA data set. Therefore, it makes sense to increasingly provide the components of the technical building equipment as ÖKOBAUDAT data sets. On a higher level, whole-building LCAs achieves a higher or more realistic informative level.

Furthermore, new prefabricated ÖKOABAUDAT data sets could be created, which would provide common composites (e.g., reinforced concrete) as a data set in order to simplify the implementation of a BIM integrated automation of whole-building LCA calculation. As soon as a component consists of several materials and thus requires different data sets with different designations for their representation, difficulties arise in the allocation of the ÖKOBAUDAT data sets to layered structures. As a possible solution and adaptation by ÖKOBAUDAT, templates for such composite components could be created based on the function "component templates" from eLCA. For example, the linking of the data sets for concrete and reinforcement steel can be provided as a new "reinforced concrete" data set.

9.5 Conclusion and Outlook

This paper describes issues and their technical improvements on the part of ÖKOBAUDAT and the IFC data format to simplify the integration and further use for the calculation of whole-building LCA according to BNB/DGNB. The adaptations presented should be seen as a basis for discussion.

Although IFC 4 provides the initial basis for the implementation of LCA information, especially environmental indicators, it can only cover parts of a whole building LCA. Therefore, further adaptations of the IFC 4 data format are useful and necessary to be conform with new requirements, e.g. DIN EN 15804: 2018.

While adaptations of the IFC data format can only be expected in the medium term through the coordination and harmonization of findings from other research projects, the adaptations presented, such as the integration of the service lives, the quality of the data sets and the completion of the life cycle modules, can be implemented in the short term to simplify whole-building LCA.

It is very important to note that this work, with the consideration of the whole-building LCA as a certification proof, is very specifically aimed at the application case of "late project phases". Application possibilities in early project phases could therefore not be considered. However, the technically shown recommendations can also be used for the implementation of an LCA in early project phases.

For example, linking through (new) LCA data sets that summaries the environmental impacts of building components or structures, e.g. exterior walls at a certain U-value, is important if no information on materiality and the exact structure is available at an early stage. Thus, in a similar way to "material mapping" in late project phases, "component mapping" in early project phases can be enabled. Valid LCA benchmarks as preliminary estimates are elementary and required for further research.

Moreover, only a small part of the research area BIM integrated LCA is affected by the quality of the data sets and their technical connections presented in this paper. Besides this and dealing with imprecise information content for the calculation of LCA in early project phases, the communication and comprehensibility of LCA results or LCA benchmarks are also decisive aspects, to support environmental optimizations more intensively in practice.

Considering all these aspects within the use of the BIM method, completely new processes, workflows and requirements for the involved stakeholders will arise, to be investigated in future research.

Acknowledgements This publication was published within the research project "Ökobilanzierung und BIM im Nachhaltigen Bauen" (SWD-F 10.08.17.7-18.29) of the Zukunft Bau program, funded by the Federal Institute for Research on Building, Urban Affairs and Spatial Development (BBSR) within the Federal Office for Building and Regional Planning (BBR).

References

Brockmann T (2019) Digitalization of building LCA and international activities—in the context of German assessment system for sustainable building. IOP Conference Series: Earth and Environmental Science. https://doi.org/10.1088/1755-1315/323/1/012108

buildingSMART International Limited (2019) Industry Foundation Classes: Version 4—Addendum 2. Pset_EnvironmentalImpactIndicators. https://standards.buildingsmart.org/IFC/RELEASE/IFC4/ADD2/HTML/link/pset_environmentalimpactindicators.htm. Accessed 5 Nov 2019

buildingSMART International Limited (2019) Industry foundation classes: version 4—Addendum 2. Pset_EnvironmentalImpactValues. https://standards.buildingsmart.org/IFC/RELEASE/IFC4/ADD2/HTML/link/pset_environmentalimpactvalues.htm. Accessed 5 Nov 2019

buildingSMART International Limited (2019) Industry Foundation Classes: Version 4—Addendum 2. IfcURIReference. https://standards.buildingsmart.org/IFC/RELEASE/IFC4/ADD2/HTML/link/ifcurireference.htm. Accessed 2 Dec 2019

Bundesministerium des Innern, für Bau und Heimat (BMI) (2020) ÖKOBAUDAT. Informationsportal Nachhaltiges Bauen. http://www.oekobaudat.de/. Accessed 12 Jan 2020

Bundesministerium für Umwelt, Naturschutz, Bau und Reaktorsicherheit (BMUB) (2019) Bewertungssystem Nachhaltiges Bauen (BNB) Neubau Büro- und Verwaltungsgebäude. Prozessqualität: Projektvorbereitung. https://www.bnb-nachhaltigesbauen.de/fileadmin/steckbriefe/verwaltungsgebaeude/neubau/v_2015/BNB_BN2015_511.pdf. Accessed 20 Dec 2019

DGNB (2018) System—Marktversion 2018. Kriterienkatalog Gebäude Neubau (3. Version)

DIN EN ISO 16739 (2017) Deutsches Institut für Normung e.V.: Industry Foundation Classes (IFC) für den Datenaustausch in der Bauindustrie und im Anlagenmanagement. Beuth Verlag GmbH

DIN EN ISO 29481-1 (2018) Deutsches Institut für Normung e.V. Bauwerksinformationsmodelle—Handbuch der Informationslieferungen. Beuth Verlag GmbH

DIN EN ISO 14040 (2009) Deutsches Institut für Normung e.V.: Umweltmanagement—Ökobilanz—Grundsätze und Rahmenbedingungen. Beuth Verlag GmbH, Berlin, Heidelberg

DIN EN 15804 (2014) Deutsches Institut für Normung e.V.: Nachhaltigkeit von Bauwerken—Umweltproduktdeklarationen: Grundregeln für die Produktkategorie Bauprodukte. Beuth Verlag GmbH, Berlin

DIN EN 15804/A2 (2018) Deutsches Institut für Normung e.V.: Nachhaltigkeit von Bauwerken—Umweltproduktdeklarationen: Grundregeln für die Produktkategorie Bauprodukte—ENTWURF. Beuth Verlag GmbH, Berlin

DIN EN 15978 (2012) Deutsches Institut für Normung e.V.: Nachhaltigkeit von Bauwerken – Bewertung der umweltbezogenen Qualität von Gebäuden – Berechnungsmethode. Beuth Verlag GmbH, Berlin, Heidelberg

DIN EN ISO 14044 (2006) Deutsches Institut für Normung e.V.: Umweltmanagement—Ökobilanz—Anforderungen und Anleitungen. Beuth Verlag GmbH, Berlin, Heidelberg

Figl H, Kerz N, Kusche O, Rössig S (2017) ÖKOBAUDAT. Basis for the building life cycle assessment. In: 2017th edn. Schriftenreihe Zukunft Bauen, Band 11. Federal Institute for Building Urban Affairs and Spatial Development within the Federal Office for Building and Regional Planning, Bonn

Gantner J, Lenz K, Horn R, Both P von, Ebertshäuser S (2018) Ökobau.dat 3.0–Quo Vadis? Buildings. https://doi.org/10.3390/buildings8090129

ISO 16739-1:2018-11 (2018) International Organization for Standardization: Industry Foundation Classes (IFC) for data sharing in the construction and facility management industries - Part 1: Data schema. Beuth Verlag GmbH

Lambertz M, Theißen S, Höper J, Wimmer R (2019) Importance of building services in ecological building assessments. E3S Web Conference. https://doi.org/10.1051/e3sconf/201911103061

Lambertz M, Wimmer R, Theißen S, Höper J, Meins-Becker A, Zibell M (2020) Ökobilanzierung und BIM im Nachhaltigen Bauen. Endbericht. Zukunft Bau (10.08.17.7-18.29). Technische Hochschule Köln (TH Köln); Bergisch Universität Wuppertal, Lehr- und Forschungsgebiet Baubetrieb und Bauwirtschaft; TMM Group. https://www.bbsr.bund.de/BBSR/DE/FP/ZB/Auftragsforschung/2NachhaltigesBauenBauqualitaet/2019/oekobilanz-bim/01-start.html?nn=436654. Accessed 4 Oct 2019

Pohl S (2014) Analyse der Rechenverfahren für die Ökobilanzierung im Bewertungssystem Nachhaltiges Bauen für Bundesgebäude (BNB). Gegenüberstellung von detailliertem und vereinfachtem Rechenverfahren. Abschlussbericht. Forschungsinitiative ZukunftBau, Band F 2911. Fraunhofer IRB Verl., Stuttgart

Rössig S (2014) eLCA Starter-Handbuch

VDI 2067 (2012) Verein Deutscher Ingenieure (VDI): Wirtschaftlichkeit gebäudetechnischer Anlagen -Grundlagen und Kostenberechnung, 1st edn. Beuth Verlag GmbH, Heidelberg

Chapter 10
Storage LCA Tool: A Tool for the Investigation of the Environmental Potential of Innovative Storage Systems in Buildings

Roberta Di Bari and Rafael Horn

Abstract With increasing energy performance requirements of buildings, new solutions and materials for thermal energy storage have been recently developed. In this regard, phase change materials (PCMs) proved to provide a good energetic performance, but their environmental potentials are still debated. In this work, a tool for environmental performance evaluation of PCM storage systems is presented. On the basis of a comprehensive collection of data coming from PCM producers, Life Cycle Assessment (LCA) analyses are carried out on material level. The integration of building energy simulations allows analyses on higher levels (component and building level). Afterwards, through comparisons among different systems the tool supports decision-making and enables optimization of PCMs storage system depending on location, building type, insulation level, and components specifications. Results proved that, despite their energy saving potentials, PCM storage systems showed not always a good environmental performance over the whole life-cycle. Main drivers are impacts from material production processes and low recycling rates. Thus, the development of such energy supply and storage concepts considering environmental life cycles has to be encouraged.

Keywords PCMs · Storage material · LCA tools · Energy storage

10.1 Introduction

The most recent trends in energy consumption and energy-related carbon emissions related to the building sector show an increasing energy use with only a limited reduction in building-related emissions. In 2018, buildings accounted for 36% of global

R. Di Bari (✉) · R. Horn
University of Stuttgart, Institute for Acoustics and Building Physics IABP, Stuttgart, Germany
e-mail: roberta.di-bari@iabp.uni-stuttgart.de

R. Horn
Fraunhofer Institute for Building Physics IBP, Stuttgart, Germany

S. Albrecht et al. (eds.), *Progress in Life Cycle Assessment 2019*,
Sustainable Production, Life Cycle Engineering and Management,
https://doi.org/10.1007/978-3-030-50519-6_10

final energy use and 39% of energy-related carbon dioxide (CO_2) emissions (Global Alliance for Buildings and Construction 2018). According to the latest information, average energy consumption in Europe will increase by 14% between 2010 and 2019 (Eurostat 2019). By considering that almost 66% of the total energy consumption in buildings is due to space conditioning and further 22% is covered by other heat uses (i.e. water heating and cooking), part of current investigations focused on strategies for energy saving.

The use of thermal energy storages (TES) offers possibilities to provide thermal energy more effectively, even generate or store electricity, both in heat or cold (Arce et al. 2011). The applied energy storage principles distinguish chemical thermal energy storage (CTES) by chemical or sorptive reactions, sensible thermal energy storage (STES) through hot water storage, and latent thermal energy storage (LTES), achievable for instance through phase change materials (PCM). The latter has been already exploited for building systems in order to obtain total energy demand savings. In this sense, it can be claimed that PCMs have an environmental improvement potential (Pielichowska and Pielichowski 2014; Aranda-Usón et al. 2013). Nevertheless, products environmental potential should consider its whole lifecycle and, in this respect, Life Cycle Assessment (LCA) analyses allow the identification of environmental performance without the risk of a shift of burdens.

Regarding the lifecycle of PCMs, some issues have arisen and recently been discussed but not yet fully resolved. In the building sector, inorganic PCMs are preferred, since they allow smaller ranges of melting points, a good conductivity, the high latent heat of fusion and their low toxicity (Liu et al. 2012). On the other hand, it is still unclear if their utilization is worthy. The benefits due the energy demand reductions should be compared with the required production and end of life impacts. Furthermore, inorganic PCMs, especially due to their high degree of sub-cooling and poor nucleating properties, can strongly decrease their storage efficiency after each cycle (Kylili and Fokaides 2016).

With the project "Storage LCA" (original "Speicher LCA"), founded by the German Federal Ministry of Economics and Technology (BMWi) and in collaboration with Fraunhofer ISE, ZAE Bayern and University of Stuttgart, comprehensive environmental assessments on storage materials and systems have been carried out with consideration of sensible, latent and thermochemical storage concepts as well as their energetic performance under different boundary conditions. In comparison with previous studies, more information about production and End-of-Life processes have been included, by collecting all relevant data directly from PCMs producers (Rubitherm Phase Change Materials PCM 2020; Rubitherm Phase Change Materials 2020). This generates more consistent and transparent LCIA results. On a higher level, PCMs should be evaluated with sufficient awareness, by including in the overall impact assessment building energy demand derived, e.g., through building energy simulations. Due to higher amount of information to be managed, seeking a tailored solutions may be not an easy and immediate action and underestimations of PCM's environmental impact can mislead final decisions.

In this paper such issues are handled with "Storage LCA Tool", conceived as an accessible and easy-to-use instrument, which supports decision making within the context of innovative energy storage concepts for building application (Di Bari et al. 2019). Investigations on thermal property are combined with LCA on different levels, from material to building level, by following a whole lifecycle approach (Fraunhofer 2019).

10.2 Goal and Scope

For "Storage LCA Tool", a modelling approach on three hierarchy levels has been established for the description of a defined system level and its evaluation (Horn et al. 2018); (Di Bari et al. 2020). The three main levels are:

- **Material**, the PCM storage material in ideal state without considering losses or degradation,
- **Component**, which is aimed at receiving and releasing thermal energy for space conditioning (heating and/or cooling) through PCM's favorable thermal storage capacity, and
- **Concept**, which in the building level, envisages the energy supply system, and include storage materials, storage components, and conventional building services systems for energy provision.

On the concept level, two main functions have been established. The first one is related to the main building specifications (net surface area, building type and insulation level, location) and defines the energy demand among the local heating and cooling period. As secondary function, thermal fluctuations in the room, which can lead to changes in the heat and/or cooling supply or provision, can be compensated.

The data platform of "Storage LCA Tool" consists of an environmental database with LCIA results for materials and installation components and energy simulation results, provided by the software TRNSYS (Thermal Energy System Specialists 2019). The two different databases interact with each other and, based on user-entered, information, environmental indicators are calculated. Storage LCA Tool allows the comparison between innovative systems and reference systems which do not entail any storage. The Life Cycle Assessment (LCA) calculations have been carried out with GaBi ts software (Thinkstep 2018) according to the standards ISO 14040 (2006) and ISO 14044 (2006). The considered indicators are climate change, expressed as GWP (Global Warming Potential, 100 years) in CO_2 eq., PENRT (Primary energy nonrenewable total), PERT (Primary energy renewable total), and PEtot (Primary energy total), expressed in MJ (see Sects. 10.3.1–10.3.2).

LCA analyses provided results for "Cradle to Grave" system boundaries, with consideration of the following modules according to DIN EN 15804 (2014) and DIN EN 15978 (2012):

- **Production**: A1–A3 Raw materials supply, transport, manufacturing
- **Use phase**: B6 Operational energy use (electricity and heating)
- **End-of-Life**: C + D Waste and credits due to recycling

Life cycle analyses consider a life cycle of 20 years in accordance with the service life of energy providing systems.

10.3 Tool Functionality

"Storage LCA Tool" aims to be a helpful instrument for decision making in the context of energy storage systems. Whether expert or not, tool users receive scientific support in the selection of environmental suitable thermal storage materials and concepts for building applications on the basis of LCA. Hereby, sensible and latent storage of materials have been analyzed, accompanied by energetic performance simulations on the building level.

Two modes, a basic and an advanced one, are distinguished (Fig. 10.1).

The *Basic Mode* enables the user to perform comparisons for storage materials at predefined working temperatures. Minimum and maximum temperatures are defined, according to the melting temperature range of the selected PCM and operating temperature of the selected distribution system. Furthermore, a simplified analysis of innovative storage system layouts on building level can be conducted and compared with a reference system. User enters information about (1) building type, (2) location, (3) insulation level, (4) energy storage and supply concept. Depending on such specifications, a list of available storage system layouts is generated and the chosen one analyzed. The final analysis is based on default storage components/combinations as well as default energetic simulations. Analogously, reference systems layouts are

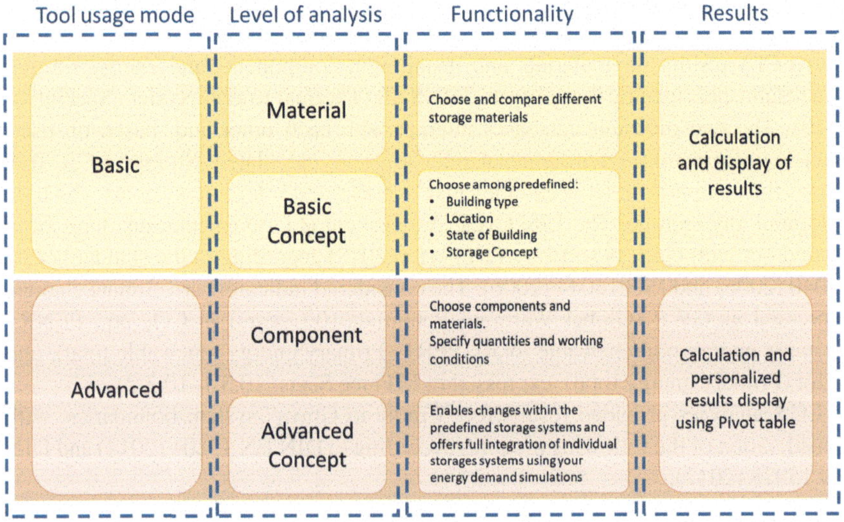

Fig. 10.1 Storage LCA Tool functionality: advanced and basic mode (Di Bari et al. 2020)

automatically generated and analyzed on the basis of LCIA and energy simulation results.

Within the *Advanced Mode*, default storage components and systems can be customized individually and analyzed. Moreover, individual energy demand simulations may be integrated and a more detailed analysis performed. To ensure consistent results, advanced mode analyses are suggested for expert users and only in the event that comprehensive information about storage system and building installations are available.

10.3.1 Analysis on Material Level

For PCMs analysis, a wide range of commercially available materials with different technology readiness level (TRL), has been considered. Those are mainly paraffin (RHXX-RTXX) and salt hydrates (SPXX) with different melting points in either encapsulated or not encapsulated variants. Generic PCM such as Generic Paraffin (see Table 10.1) have been included as well (Rubitherm Phase Change Materials PCM 2020; Rubitherm Phase Change Materials 2020; Di Bari et al. 2019). The energy density is calculated by taking into account the PCMs' physical properties and working temperatures dictated by the distribution system.

The environmental impacts (GWP, PENRT, PERT, PET) for each material are calculated in an LCA using the GaBi ts database (Thinkstep 2018). The selected functional unit is kWh (stored energy) which allows analyses depending on the energy storage density (E_d). A unit transformation is carried out and the final LCIA is established as shown in the Eq. (1).

Table 10.1 PCM materials list, divided in paraffin and salt hydrate (Rubitherm Phase Change Materials PCM 2020; Rubitherm Phase Change Materials 2020; Di Bari et al. 2019)

PCM	
RH10HC (Encapsulated or not)	Paraffin
RH11HC (Encapsulated or not)	
RH18 (Encapsulated or not)	
RT18HC (Encapsulated or not)	
RT21 (Encapsulated or not)	
RT24 (Encapsulated or not)	
RT62HC (Encapsulated or not)	
Generic Parafin	
SP21 EK (Encapsulated or not)	Salt hydrate
SP58 (Encapsulated or not	
Sodium acetate (Encapsulated or not)	
SP15 (Encapsulated or not)	

$$\mathrm{GWP}\left[\frac{\mathrm{kgCO_2\,eq.}}{\mathrm{kWh}}\right]=\mathrm{GWP}\left[\frac{\mathrm{kgCO_2\,eq.}}{\mathrm{kg}}\right]\cdot\frac{1}{\mathrm{E_d}\left[\frac{\mathrm{kWh}}{\mathrm{kg}}\right]},\tag{1}$$

10.3.2 Analysis on Components and System Level

On component level, functional units should be established depending on each element. By considering, for instance, storage containment, storage volume [m^3] is recommended as functional unit; a gas boiler can be analyzed with reference to its nominal power output [W].

The reference and storage systems included in "Storage LCA Tool" are shown in Table 10.2. Those are grouped in centralized heating system, centralized cooling and decentralized (for both heating and cooling system) (Di Bari et al. 2019). For each of them different energy distribution pathways are considered (e.g., radiators and underfloor heating).

On building level, energy simulations have been carried out for different climate zones expressed by a specific location (Athens, Strasbourg or Helsinki), as well as the building type (office block, apartment building or single family house) in which different insulation levels have been provided (Table 10.3) (Di Bari et al. 2019, 2020).

Table 10.2 List with energy supply concepts with components belonging to storage and distribution systems

Supply concept	Centralised heating system	Centralised cooling system	Decentralised heating and cooling system
Reference system	Gas boiler	Split device water chiller	Water chiller
Storage system	HWT + ST + Gas PCM + ST + Gas	Water chiller + PCM Water chiller + CWT	PCM surface cooling + water chiller PCM-Ventilation systems
Distribution systems	Radiators Underfloor heating	Surface cooling Fan coil	Air
Additional system specifications	Storage material Storage volume Solar collector type + field	Storage material Water-Chiller power Storage volume	Storage material Water-Chiller Power Mass Distribution Storage volume

System specifications are entered by users Di Bari et al. (2019)

Table 10.3 List of available locations, building types and insulation levels (Di Bari et al. 2019)

Location	Helsinki	Strasbourg	Athens
Building type	Detached House Apartment building Office block		
Insulation level	Moderate insulation Energy efficient building	Little insulation Moderate insulation Energy efficient building	No insulation Little insulation Moderate insulation Energy efficient building

10.3.3 "Storage LCA Tool" Functionality

With the selection of building analyses, the user is addressed to a spreadsheet composed by a mask and a navigator (see Fig. 10.2): the navigator on the right guides the user within the analysis. The user guide accompanies the tool. At any time, the user has the possibility to get help by clicking on the Button "Info user guide" (in each spread sheet on the left) (Di Bari et al. 2019).

The selection mask (on the center) allows the selection of a building type, location, and energetic state of the building which expresses a building insulation level. All necessary information is entered by drop down list.

Within the Basic Mode, no additional specifications are required. However, within Advanced Mode, material specifications and quantities regarding installation sets are necessary. Results are provided in both numerical and graphical forms and resumed in a pre-printed form.

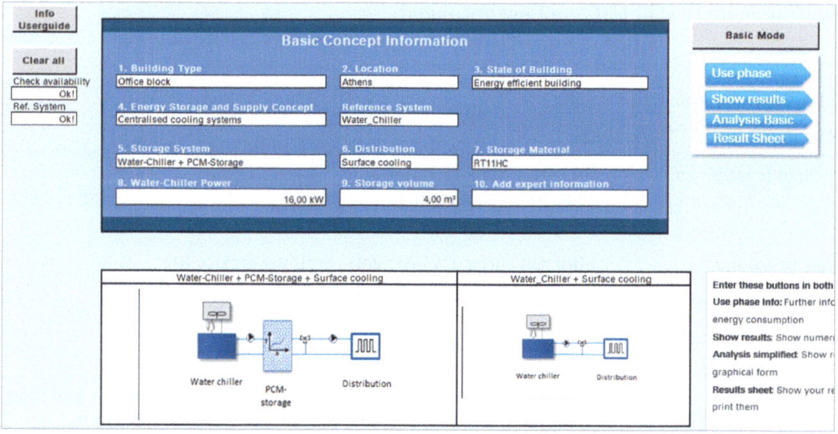

Fig. 10.2 Storage LCA tool: main and navigation masks (Di Bari et al. 2019)

10.4 Storage System Assessment Results

In this section, the results of two exemplary analyses are presented, one cooling and one heating innovative energy storage concept, for which system specifications and simulations results were predefined ("Basic" tool usage mode). Table 10.4 resumes details of LCA analyses and provides the layout of the selected case studies.

The first energy storage concept has been analyzed in a Mediterranean climate area, where building energy demand for space conditioning may be due to cooling. On the contrary, in the Northern-European area an innovative heating system has been analyzed. For both examples, the same building type and energy standard have been chosen, namely an energy efficient office block. For a more complete overview on PCM material level, two different groups have been selected, a Paraffin (RT11HC) and a Salt hydrate (SP58), respectively.

In a first instance, analyses are carried out on the whole building level. Through the comparison between innovative storage systems and conventional ones provided by "Storage LCA Tool", it will verify the worthiness of PCM storage in terms of total environmental impacts. Lastly, the generated graphical and numerical results can identify shortcomings and address optimization possibilities in order to guarantee an efficient energy storage and less environmental impacts.

10.4.1 Cooling System

The selected cooling system consists of a water-chiller + PCM storage of 8 m^3 volume and cooling surface as distribution system. The selected PCM material is a paraffin, namely RT11HC (not encapsulated).

Table 10.4 LCA analyses details

Specification of LCA Analyses	
System boundaries	A1–A3: Manufacturing B6: Energy consumption C + D: End of life and credits
Functional unit	[m^2 net surface*y] 20 years' building service life
LCIA categories	Global Warming Potential (GWP) [kg CO_2 eq.] CML 2001—January 2016 PENRT PERT
Case study: cooling system	Office Block—Athens—Moderate insulation Water-chiller + RT11HC storage (8.00 m^3 volume) + Cooling surface
Case study: heating system	Office Block—Helsinki—Moderate insulation Gas Boiler + Solar collector (101.8 m^2) + SP58 storage 15.50 m^3 volume) + Radiators

The LCA analyses are carried out and compared with a reference system made of a Water-Chiller with cooling surface with equal characteristics. According to the results, the innovative system accounts total GWP of 10.68 kg CO_2 eq./m^2y. This result is slightly higher in comparison with the reference system (9.99 kg CO_2 eq./m^2y). In terms of total primary energy (PEtot), the storage seems to be advantageous and, as shown by Fig. 10.3, leads to energy savings up to 35%.

A following analysis considers components belonging to the storage and energy supply systems. Those are PCM storage materials, a High-density polyethylene (HDPE) containment (whereas PCM is not encapsulated) and an insulation made by extruded polystyrene (XPS). Further components are stainless steel valves and a heat exchanger made by polypropylene (PP) capillary tubes. All the amounts were assigned to the building energy simulations and uploaded in "Storage LCA Tool" data platform. The interconnection between energy simulation and LCIA databases leads to the final results which are presented in the following. In a first instance, the whole storage system is analyzed as a whole, without considering impacts due to the use phase. As shown by Fig. 10.4, most of the GWP impacts are due to the production phase while the End-of-Life (EoL) of the storage system have a positive value. This is due to the low recycling rate and wasteful processes of the defined materials.

In a second instance, an investigation of the single component is carried out. The GWP over the storage system lifecycle is summed up and the contributions of each component are distinguished. As a result (see Fig. 10.5), the selected PCM has the highest GWP share (76%).

To recap, the results of the analysis proved that, for the selected case, PCM storage system is not advantageous. Nevertheless, a revision of the entered specifications is suggested in order to derive better outcomes. The significant share of PCM material could be reduced, for instance, by decreasing the storage volume.

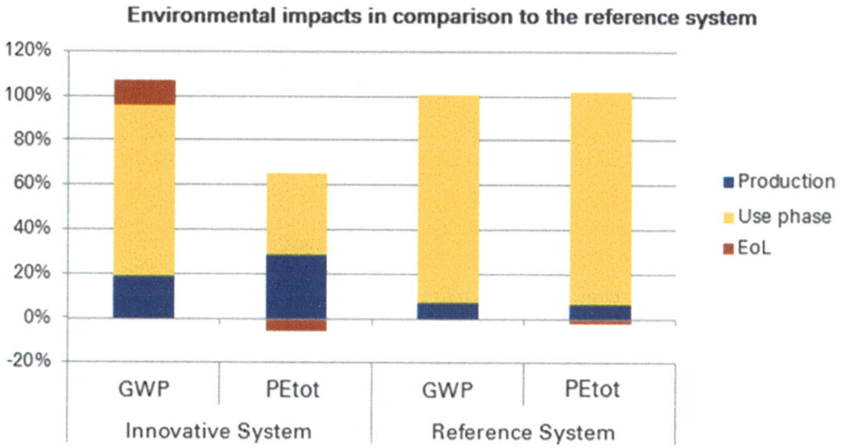

Fig. 10.3 Comparison between Innovative system (Water chiller + RT11HC 8.00 m^3 storage volume + cooling surface) and Reference system (Water chiller + cooling surface). Case study: energy efficient office block located in Athens (Di Bari et al. 2019)

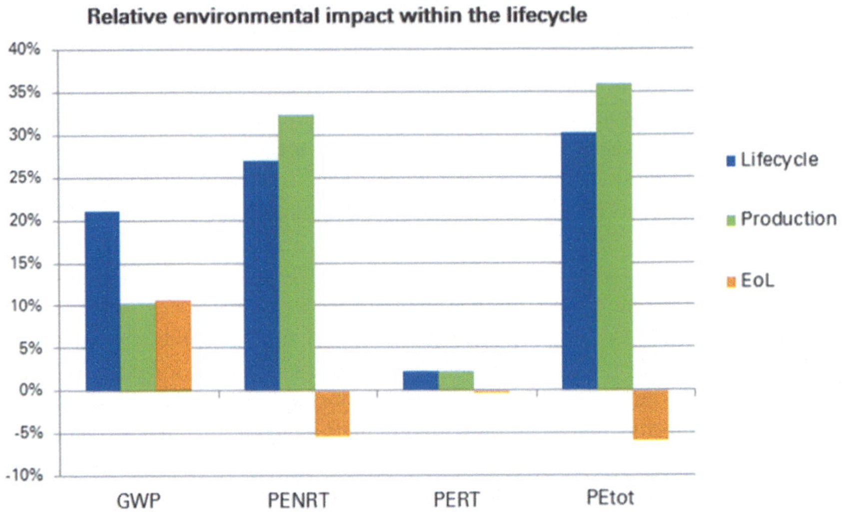

Fig. 10.4 Relative environmental impacts of Innovative RT11HC (8.00 m^3) storage systems (Di Bari et al. 2019)

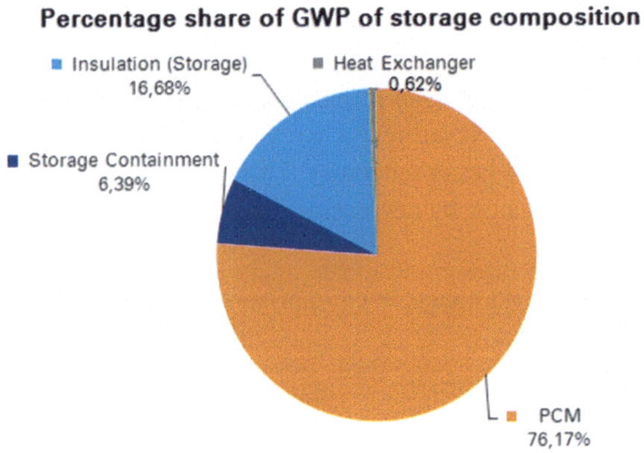

Fig. 10.5 Analysis on component level of Innovative RT11HC (8.00 m^3) storage system (Di Bari et al. 2019)

The same energy concept is therefore provided with a storage containment which allows 4 m^3 volume. This modification has been easily performed through the alternative selection from the storage volume drop-down list. As shown by tool final graphs (Figs. 10.6, 10.7), this choice proved to be effective. The total GWP sinks by up to 9.54 kg CO_2 eq./m^2 y without relevant effects for the storage potential. With regards to GWP, savings of 5% are assessed, while energy savings are almost 42%. In conclusion, the selected PCM - energy storage layout, differently from the previous one, proved to have environmental potentials.

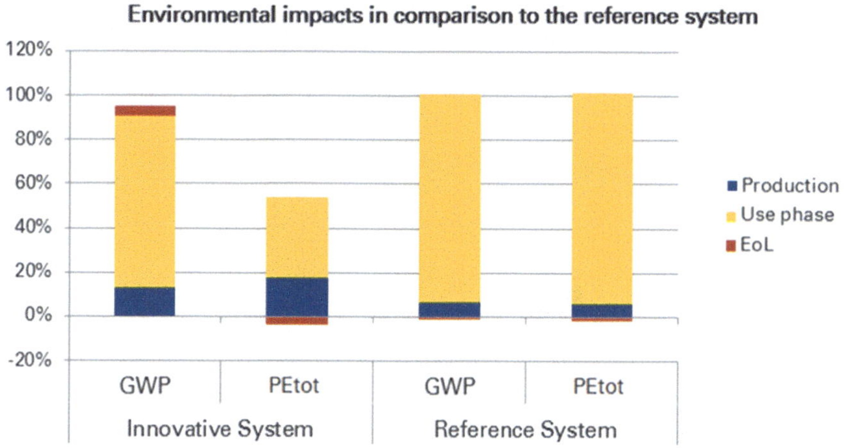

Fig. 10.6 Comparison between Innovative system (Water chiller + RT11HC 4.00 m^3 storage + Cooling surface) and Reference system. Case study: Energy efficient office block located in Athens (Di Bari et al. 2019)

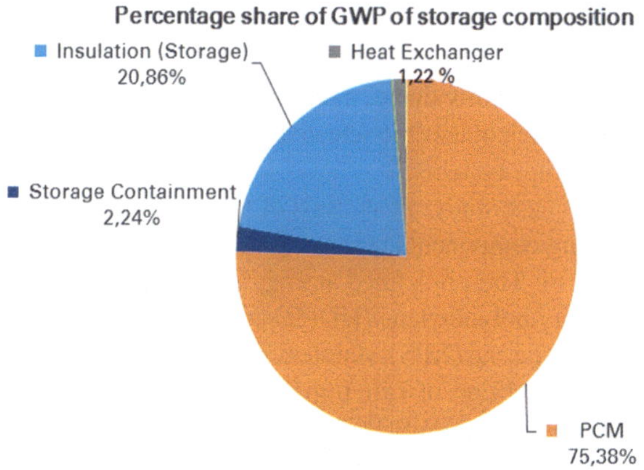

Fig. 10.7 Analysis on component level of Innovative RT11HC (4.00 m^3) storage system (Di Bari et al. 2019)

10.4.2 Heating System

For analyses on heating systems, a Gas Boiler + Solar collector (flat plate with 101.8 m^2 surface) has been selected with PCM storage of 15.50 m^3 volume. The chosen distribution systems are radiators. The storage material is a not encapsulated salts hydrated, namely SP58. The reference system is made by a Gas Boiler with radiators and does not account solar collectors.

The comparison carried out by "Storage LCA Tool" between them is in favor of the PCM storage system. According to the results, the innovative system produces

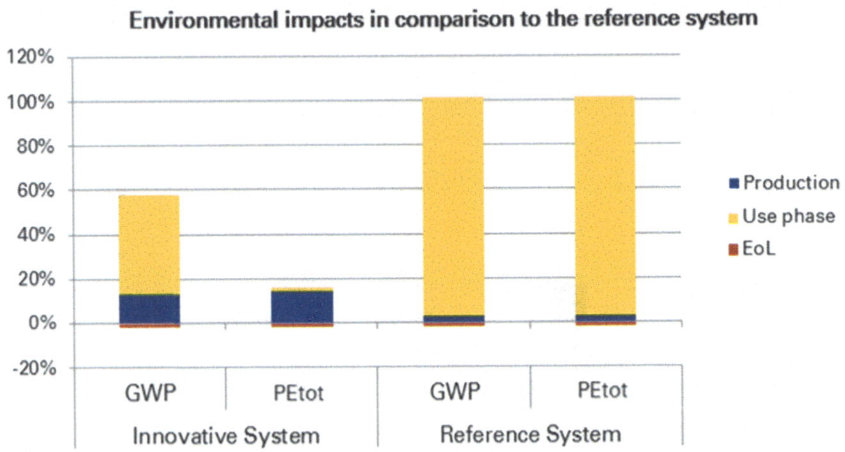

Fig. 10.8 Comparison between innovative system (Gas Boiler + Solar collector + SP57 15.5 m^3 storage + Radiators) and reference system (Gas Boiler + Radiators). Case study: Energy efficient office block located in Helsinki (Di Bari et al. 2019)

totally 24.58 kg CO_2 eq./m^2y which is almost 50% less than the GWP calculated for the reference system (43.62 kg CO_2 eq./m^2y). Energy saving are significant as well (see Fig. 10.8). This is certainly due to the effective combination of solar collectors and storage systems, which makes the selected solution advantageous in comparison with conventional heating systems.

As for the previous case study, an analysis of the storage system has been carried out, by not considering components belonging to the heating (e.g., solar collectors) and distribution system. The predefined storage system includes SP15 (not encapsulated), a High-density polyethylene (HDPE) containment, containment insulation made by extruded polystyrene (XPS), stainless steel valves and a capillary tubes heat exchanger. Production and End-of-Life impacts have been summed up and grouped by component. As well as for GWP, PCM have the highest share (75%) of primary energy not renewable (PENRT). This is due to the wasteful production processes which still require high demand on energy, coming mostly from fuel fossils (see Fig. 10.9).

10.5 Discussion and Outlooks

In this work, applications of the "Storage LCA Tool" have been conducted. Among its several functionalities, the comparison of energy concepts enables faster and easier decisions in the field of PCM applications in the building sector.

Differently from previous works, LCA analyses can be provided automatically on different levels, hereby allowing more tailored solutions for energy savings and environmental impacts reductions.

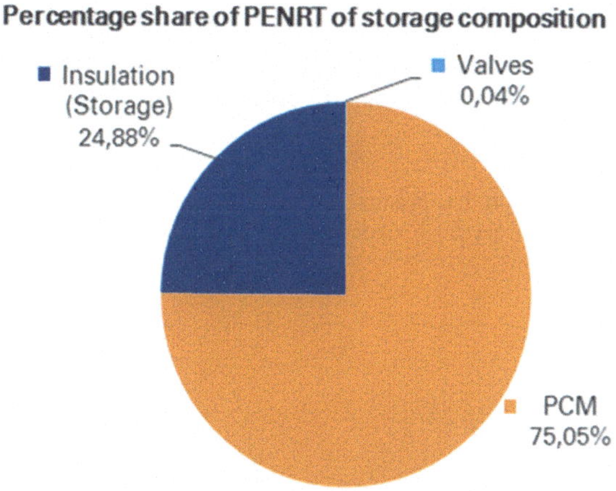

Fig. 10.9 Analysis on component level of Innovative SP58 storage system (Di Bari et al. 2019)

With regard to this, the integration of updated information coming from PCM producers proved to be relevant for the assessment of the environmental potential of storage systems. Outcomes highlighted that paraffins are still advantageous in terms of thermal storage but their production requires higher quantities of primary energy and produces higher impacts (Horn et al. 2018). In addition to this, despite all the developments and efforts of PCM producers, a favorable recycling potential cannot yet be recorded. This calls for more research on organic paraffins, which is nowadays still ongoing and may represent a better solution for the reduction of PCMs environmental impacts.

With reference to the current available storage systems, the results of this work proved advantages in terms of energy savings for innovative storage systems especially if combined with sources of renewable energy. On the contrary, environmental advantages cannot always be ensured over the building lifecycle. The building type, the location, and the insulation level can dictate the energy demand and the selected storage system should take this into account. PCM storage components (e.g., storage containments), materials and quantities should be chosen in order to optimize the overall energy performance and avoid adverse extra environmental impacts. As in our examples, the accurate selection of the storage volume proved to be relevant for the final results.

This procedure was carried out immediately through "Storage LCA Tool", for which further technical improvements are possible. An extension of LCA system boundaries, through the inclusion, meaning refurbishment and substitution of components (B2–B4 Module according to EN15804) can provide more accurate results and enables the extension the analysis duration up to 50 years. Furthermore, more environmental indicators can enhance the final evaluation.

Lastly, it is necessary to enrich the data platform of "Storage LCA Tool" with new materials, storage components, and building energy simulation results and derive overall evaluations of the environmental value of storage systems. This will be carried out by an auxiliary tool aimed to gather all results and provide automatically a generalizable analysis by varying parameters and comparing automatically all PCM storage systems not only with reference systems, but also with other solutions, such as water storage.

Acknowledgements This work has been created as part of the project "Speicher LCA" funded by the German Federal Ministry for Economic Affairs and Energy under grant agreement FKZ 03ET1333C.

References

Aranda-Usón A, Ferreira G, López-Sabirón AM et al (2013) Phase change material applications in buildings: an environmental assessment for some Spanish climate severities. Sci Total Environ 444:16–25. https://doi.org/10.1016/j.scitotenv.2012.11.012

Arce P, Medrano M, Gil A et al (2011) Overview of thermal energy storage (TES) potential energy savings and climate change mitigation in Spain and Europe. Appl Energy 88(8):2764–2774. https://doi.org/10.1016/j.apenergy.2011.01.067

Di Bari R, Horn R, Nienborg B et al (2019) Speicher-LCA Tool. University of Stuttgart, Department of Life Cycle Engineering (GaBi), Stuttgart, Germany

Di Bari R, Horn R, Nienborg B et al (2020) The Environmental Potential of Phase Change Materials in Building Applications. A Multiple Case Investigation Based on Life Cycle Assessment and Building Simulation. Energies 13(12):3045.https://doi.org/10.3390/en13123045

DIN EN 15804 (2014) European committee for standardization sustainability of construction works environmental product declarations. Core rules for the product category of construction products: (German version). https://www.cen.eu

DIN EN 15978 (2012) European committee for standardization sustainability of construction works. Assessment of environmental performance of buildings. Calculation method (German Version). https://www.cen.eu/. Accessed 16 Apr 2020

DIN EN ISO 14040 (2006) International standardization organization environmental management—life cycle assessment—principles and framework. https://www.iso.org. Accessed 16 Apr 2020

DIN EN ISO 14044 (2006) International standardization organization environmental management—life cycle assessment—requirements and guidelines. https://www.iso.org. Accessed 16 Apr 2020

Eurostat (2019) Energy saving statistics: Distance to 2020 and 2030 targets for primary energy consumption, EU-28

Fraunhofer ISE (2019) Ökologische Bewertung ausgewählter Konzepte und Materialien zur Wärme-und Kältespeicherung (Speicher-LCA). https://www.ise.fraunhofer.de/de/forschungsprojekte/speicher-lca.html. Accessed 16 Apr 2020

Global Alliance for Buildings and Construction (2018) Global 2018 status report: towards a zero-emission, efficient and resilient buildings and construction sector. https://www.unenvironment.org/resources/report/global-status-report-2018

Horn R, Burr M, Fröhlich D et al (2018) Life cycle assessment of innovative materials for thermal energy storage in buildings. Procedia CIRP 69:206–211. https://doi.org/10.1016/j.procir.2017.11.095

Kylili A, Fokaides PA (2016) Life cycle assessment (LCA) of phase change materials (PCMs) for building applications: a review. J Build Eng 6:133–143. https://doi.org/10.1016/j.jobe.2016.02.008

Liu M, Saman W, Bruno F (2012) Review on storage materials and thermal performance enhancement techniques for high temperature phase change thermal storage systems. Renew Sustain Energy Rev 16(4):2118–2132. https://doi.org/10.1016/j.rser.2012.01.020

Pielichowska K, Pielichowski K (2014) Phase change materials for thermal energy storage. Prog Mater Sci 65:67–123. https://doi.org/10.1016/j.pmatsci.2014.03.005

Rubitherm Phase Change Materials PCM (2020) SP-Serie. https://www.rubitherm.eu/index.php/produktkategorie/anorganische-pcm-sp. Accessed 16 Apr 2020

Rubitherm Phase Change Materials PCM RT-Serie (2020) https://www.rubitherm.eu/index.php/produktkategorie/organische-pcm-rt. Accessed 16 Apr 2020

Thermal Energy System Specialists, LLC (2019) TRNSYS: transient system simulation tool, Madison, WI, USA

Thinkstep ASC (2018) GaBi ts Professional, Chicago, IL, USA

Chapter 11
Economic and Environmental Optimization of Rotary Heat Exchangers: A Closer Look at the Conflict

Eloy Melian, Harald Klein, and Nikolaus Thißen

Abstract Rotary heat exchangers belong to the most efficient heat exchangers for gas streams. This technology is commonly used in power plants, paint shops and heating, ventilation and air conditioning systems. Yet an optimization taking into account both economic and ecological aspects is absent in literature. This work uses a simulation model developed at the Pforzheim University of Applied Sciences, combined with LCA and economic data to optimize rotors. From an environmental perspective, the use phase is where the bigger impacts take place. In the economical side, the use phase is also the most important. For the optimization, a commercial software for this purpose is used. The results show a conflict between the two possible optimization goals. Additionally data of two commercially available rotary heat exchangers from the Eurovent Database are also compared to the results from optimization via simulation and there is evidence that in some specific cases an improvement in the environmental or economical aspect can be done without compromising the other aspect.

Keywords Rotary heat exchanger · Thermal wheel · Optimization

---

E. Melian (✉) · N. Thißen
Institute for Industrial Ecology (INEC), Pforzheim University of Applied Sciences, Tiefenbronnerstr. 65, 75175 Pforzheim, Germany
e-mail: eloy.melian@hs-pforzheim.de

N. Thißen
e-mail: nikolaus.thissen@hs-pforzheim.de

H. Klein
Technical University of Munich, Bolzmannstr. 15, 85745 Garching, Germany
e-mail: harald.klein@tum.de

S. Albrecht et al. (eds.), *Progress in Life Cycle Assessment 2019*,
Sustainable Production, Life Cycle Engineering and Management,
https://doi.org/10.1007/978-3-030-50519-6_11

11.1 Introduction

In gas streams, most types of heat exchangers show low efficiencies (VDI-Fachbereich Technische Gebäudeausrüstung 1997). Rotary heat exchangers on the opposite have proved to be on the most efficient range for gas to gas heat exchange (O'Connor et al. 2016). However, its design has not changed since the 1960s (Hochschule Pforzheim 2019). The purpose of this work is to optimize the rotor design parameters and operating conditions based on Life Cycle Assessment (LCA) and on economic considerations.

Information on LCA for a complete rotary heat exchangers is nonexistent in current literature. Economic assessment is usually provided by manufacturers and also very limited in content. An optimization of rotary heat exchangers considering ecological and economic goals has not been found in literature.

11.1.1 Rotary Heat Exchangers

Rotary heat exchangers are also known in literature as "thermal wheel", "heat recovery wheel", "Kyoto wheel" or "rotors". For convenience, in this work rotary heat exchangers will be referred to as "thermal wheels" or "rotors".

Rotors are made out of metal sheets and their appearance resembles coiled corrugated fiberboard as shown in Fig. 11.1.

As rotors recover heat in gas streams this technology is used commonly in power plants, paint shops as well as heating ventilation and air conditioning (HVAC). In the latter applications, rotors are often made of aluminum.

Rotors are placed into a housing to exchange energy between two gas streams as shown in Fig. 11.2. This housing separates both gas streams and each half of the

Fig. 11.1 Rotor appearance

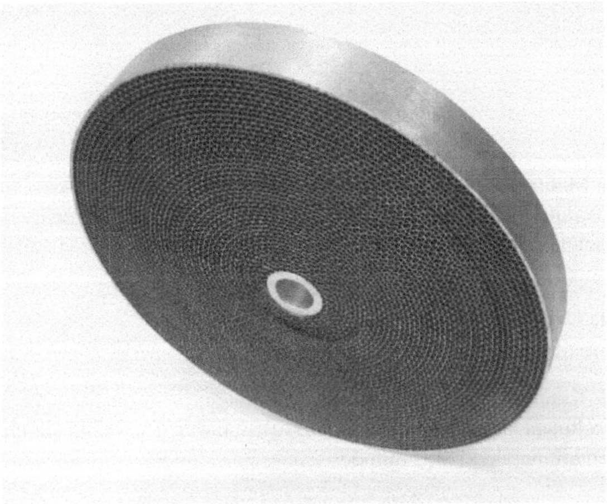

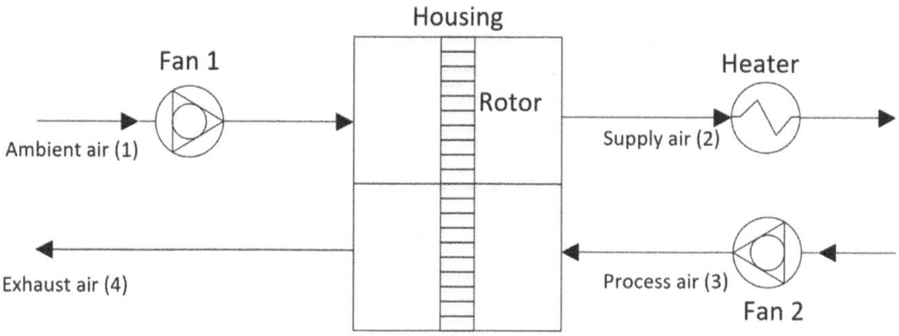

Fig. 11.2 Rotor scheme in a process

rotor is in a different side of the housing. For example, in winter, when the rotor is on steady state conditions, the ambient air (1) is blown by a fan into the housing. In the housing, the air flows through the channels of the rotor structure and takes the heat from the rotor. The air leaves the rotor preheated as supply air (2). Since the supply air might not be at the required temperature for a downstream process or building, the supply air is heated by a heater to the required temperature.

Simultaneously, the process air that needs to be taken out of the process or building system flows through the rotor structure and heats the metal sheets. Finally the exhaust air (4) leaves the system. Since the two halves of the rotor are in different sides of the housing, and the thermal wheel is continuously rotating, one half is continuously storing energy while the other is releasing it.

The performance of rotary heat exchangers depends on the constructive and operational parameters for which it is designed. Among the most important, one can find the empty tube gas velocity v_a f the air and rotor length l both shown on Fig. 11.3), material thickness of the metal sheets, wave inclination angle (or wave length, which is analogous) and the wave height of the corrugated metal sheets (shown on Fig. 11.4).

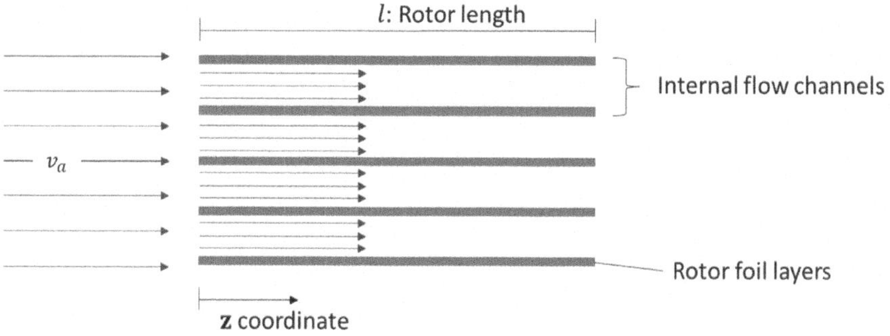

Fig. 11.3 Rotor length and empty tube gas velocity

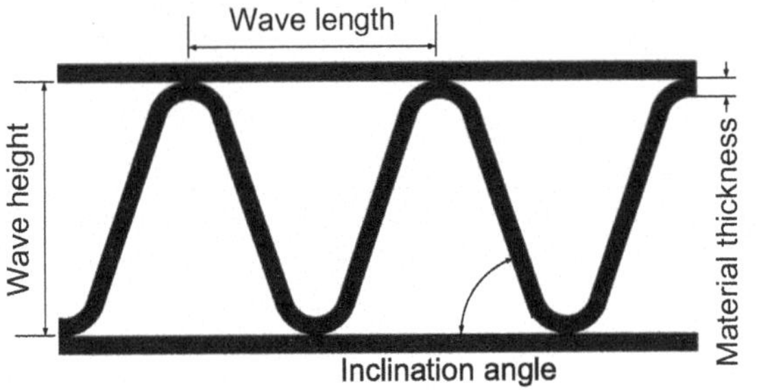

Fig. 11.4 Corrugated metal sheet parameters

These are the parameters that mainly affect the energy recovery efficiency from the rotor. The rotor is part of the system for which the required heat $\dot{Q}_{required}$ can be calculated by:

$$\dot{Q}_{required} = \dot{Q}_{rotor} + \dot{Q}_{heater} \tag{11.1}$$

Consequently, by a given $\dot{Q}_{required}$ for reaching a desired temperature, the recovered heat from the rotor $\dot{Q}_{rotor}$ also affects the energy needed at the heater downstream $\dot{Q}_{heater}$ (also shown in Fig. 11.2. The more energy the rotor recovers, the less additional energy is needed at the heater. These parameters also influence the pressure drop by the rotor and therefore the power needed by the fans (Fig. 11.2). For example, a smaller wave height could be beneficial for energy recovery purposes, but this would increase the pressure drop of the rotor and thus the power consumption from the fans.

11.2 Methods

11.2.1 Simulation

A simulation model that takes into consideration the thermodynamic and fluid dynamics for a rotor was developed within the project SEROW at the Pforzheim University of Applied Sciences. This simulation takes as input the constructive and operational parameters of rotors and was proved to be accurate on predicting the heat recovery efficiency at different operating conditions in the pilot plant (Hochschule Pforzheim 2019). Results from simulations show that it is able to successfully predict the thermal efficiency of other rotary heat exchangers tested by Eurovent (Eurovent Certita Certification Certified products 2019) too.

11.2.2 General Considerations

Based on results from simulation, ecological and economical assessment can be done. The general assumptions for these assessments are listed in the following Table 11.1.

Usually rotors in HVAC uses can have a lifetime of 20 years. In the paint shops this lifetime is reduced to 10 years. Therefore, the more conservative value is used as an extreme case. In the cases where the rotors are used for HVAC purposes, for the same constructive and operational parameters, rotors are more economically attractive because of longer product life. Regarding the temperature difference between the process and the ambient air, a 10 °C difference is assumed. This value highly depends on the climate at the location where the rotor is installed.

Another assumption for the study is that the process downstream had no leaking, so that the mass flow in both gas streams (ambient air and process air) of the rotor are equal. For the LCA and economical assessment, a rotor with typical constructive parameters was selected from the Eurovent database (Eurovent Certita Certification Certified products 2019) (Table 11.2).

The calculation of the additional power consumption of the fan due to the rotor pressure drop is incorporated in the simulation model. The total fan efficiency (fan, driver and transmission) is assumed to be 65%, which is a regular efficiency by a back-curved-blade centrifugal blower (Perry et al. 1997). For the heater, it is assumed that

Table 11.1 General assumptions for economical and life-cycle assessments

Parameter	Value
Functional unit (m^3 fresh air/h)	10,000
Rotor lifetime (years)	10
Working days (d/year)	250
Working hours (h/d)	8
Ambient air (1) temperature in °C	10
Process air (3) temperature in °C	20

Table 11.2 Rotor values (Eurovent Certita Certification Certified products 2019)

Parameters	
Manufacturer	Lautner Energiespartechnik GmbH
Model	P_17-1100-WZV
Material thickness	60 μm
Wave height	1.4 mm
Length 10 waves	3,8 mm
Rotor length	20 cm
Rotation velocity	14 rpm
Material	Aluminum

it is a natural gas furnace for industrial purposes (>100 kW), with the corresponding efficiency (1 MJ heat per 1.0526 MJ fuel) from the Ecoinvent data (ecoinvent 2018).

In the following chapters an economical assessment and Life Cycle Assessment (LCA) is done. Therefore the simulations are compared to a base case where no heat recovery (no rotor) is used. In this reference case the pressure drop of the rotor is considered zero and the supply air (2) is heated only by the heater (see Fig. 11.2).

11.2.3 Economical Assessment

The main objective in this section is to understand how profitable rotors are. Different rotor constructions would have different investment and operating costs. Therefore, the main parameter to be assessed is the Internal Rate of Return (IRR).

Investment. The rotor investment costs are considered based on the work from Pufal (2017). With his work, it is able to estimate the complete rotor market prices based on the weight of the aluminum used.

Operating costs. The operation costs are divided in two categories: electricity and heat costs. Electricity costs are based on the power consumption of two fans for ambient and process air. This power is calculated only to compensate the pressure drop caused by the rotor. The empty tube gas velocity of the ambient air is 3 m/s, which is a common industrial value and the most economical from the 1–3 m/s range presented by Eurovent database. The power consumption due to the motor of the rotor or the control unit is neglected, since it is some orders of magnitude smaller than the fan power. The electricity costs are based on the assumption that the electricity price is 0.30 €/(kW h) (GET Sol 1 GmbH Strompreis pro 2019).

The other category is the heat costs from an assumed natural gas boiler. Since the rotor recovers heat to the supply air, the energy requirements at the heater are lower for reaching the desired process temperature. Therefore, the rotor saves energy from the heater. For the economic assessment, the heat provided by the rotor is considered as heat savings. The price for the gas is assumed to be 0.06 €/(kW h) based on the current market values (Anondi GmbH Was kostet eine Kilowattstunde Gas? 2019).

Recycling. Since the current aluminum scrap price is around 1.70 €/kg (Transfactor Technologies SRL 2019), this costs are not significant compared to the costs of a rotor or the operating costs (Pufal 2017) and are therefore not considered in the economical assessment.

Finally the Internal Rate of Return (IRR) is calculated. As investment, the investment price of the rotor is considered based on previous work (Pufal 2017). The cash flow is the energy savings minus the fan electricity expenses.

11.2.4 Life Cycle Assessment

Data for the assessment is taken out of the database Ecoinvent 3 with a cut-off system model (ecoinvent 2018). Following the same methodology as Pufal (2017), GWP 100 is the only category assessed in this study, since it is currently a leading indicator for climate change (Table 11.3).

Manufacturing. The manufacturing process of the rotor is assessed based on the work of Pufal (2017) within the SEROW Project at the Pforzheim University. This work was in cooperation with a rotor manufacturer and all the steps for the production of a rotor are analyzed in detail. These include the transformation of aluminum from bauxite to metal sheets, the transportation of the metal sheets to the rotor manufacturing site, the rotor production and the delivery to the final customer. Therefore, the energy consumption and material use in this life cycle phase can be calculated. Since the demand of rotary heat exchangers takes place around Europe, the distance to the customer is assumed to be 1000 km (Table 11.4).

Use Phase. During the use phase, the main electricity consumption is due to the power of the fans, for which the empty tube gas velocity of the ambient air is assumed to be 3 m/s. The energy consumption due to the motor of the rotor and the control unit is neglected, since it is together at least one order of magnitude smaller than the electrical fan power. The other point in the system where environmental impact takes place are the CO_2 emissions because of the natural gas boiler acting as a heater (Fig. 11.2). Due to the rotor, the natural gas requirements and emissions are reduced.

Recycling. The aluminum recycling rate in construction waste, like rotors, is over 70% (Graedel et al. 2015). Furthermore, in industrial processes closed cycles are typical (Graedel et al. 2011). Therefore, it was assumed that 90% of the total metal (aluminum for the rotor and steel) are successfully recycled, where at the end of the recycling phase secondary metals are produced. As usual in LCA, the

Table 11.3 Used values for LCA for transportation, electricity and gas (ecoinvent 2018)

	GWP 100	
Transportation by Lorry	0.1706	kg CO_2/(tonne·km)
Gas	0.0696	kg CO_2/MJ
Electricity	0.6476	kg CO_2/kWh

Table 11.4 GWP values in the manufacturing process (ecoinvent 2018)

	GWP 100	
Aluminum ingot	19.9368	kg CO_2/kg
Polystone M	1.9229	kg CO_2/kg
Steel	2.552	kg CO_2/kg
Electric motor	4.4325	kg CO_2/kg
Metal sheet production	0.6108	kg CO_2/kg

Table 11.5 GWP values in the recycling process (ecoinvent 2018)

	GWP 100	
Aluminum recycling	0.72974	kg CO_2/kg
Polystone M recycling	2.569	kg CO_2/kg
Steel recycling	0.6684	kg CO_2/kg

recycling process substitutes the primary metal production (Dubreuil et al. 2010). Consequently, the CO_2 emissions from the primary production in the manufacturing phase are partially saved during the recycling phase. Hence, the recycling process saves energy (and CO_2 emissions) compared to primary production. For the plastic, 90% of it is thermally used (Table 11.5).

11.2.5 Optimization Considerations

An optimization of rotors can be done using the data from the previous sections. In one case, the optimization goal is the reduction of the GWP emissions, and in the other case, the maximization of the Internal Rate of Return (IRR). The simulation software developed within the SEROW Project is used working together with the commercially available OptQuest (OPTTEK SYSTEMS I …) optimization engine. In both cases, the optimization is subject to constraints concerning the design and operation of the rotor (see Table 11.6).

Traditional rotors have been built with a rotor length between 20 and 40 cm. Within the SEROW Project there was evidence that this range was not the optimum under economical or environmental aspects (Hochschule Pforzheim 2019). The wave height and empty tube velocity within the ranges shown in Table 11.6 have an optimum that is mutually dependent [8]. The inclination angle is limited to this range shown in Table 11.6 because of the validity of the model used from the project SEROW. The optimization is based on the general considerations from the simulation Sect. (11.2.2) regarding temperature, functional unit, lifetime and working hours. Additionally, the material thickness is assumed to be 60 μm, since this is the thinnest sheet used in the industry. Thicker metal sheets are less economical and provide less energy recovery (Hochschule Pforzheim 2019) and therefore less CO_2 reductions. Another

Table 11.6 Optimization constraints

Parameter	Min	Max
Rotor length in cm	15	400
Wave height in mm	1	4
Inclination angle in °	30	60
Empty tube gas velocity ambient air in m/s	1	4

Table 11.7 Manufactured rotor comparison (Eurovent Certita Certification Certified products 2019)

Parameters		
Manufacturer	Lautner Energiespartechnik GmbH	KASTT, spol. s r.o
Model	P_17-1100-WZV	ROV UNI/3000-T-1,4
Material thickness in μm	60	70
Wave height in mm	1.4	1.4
Length 10 waves in mm	3.8	3.7
Rotor length in cm	20	20
Rotation velocity in rpm	14	12
Material	Aluminum	Aluminum

operational parameter is the rotational speed of the rotor which was set to 24 rpm, which is a common industry value. At a constant rotational speed, a rotor with a longer rotor length would carry over more exhaust air into the supply air. To compensate this carry over effect, longer rotors have to be dimensioned with a higher surface area and therefore diameter, so the total volume of the heated supply air minus the carry over gas is still the functional unit of fresh air of 10 000 m^3/h, which is a middle value for one-piece rotors (Pufal 2017) (Table 11.7).

Two typical commercial rotors from the Eurovent database (Eurovent Certita Certification Certified products 2019) are additionally compared to the results of the optimization calculations in terms of GWP savings over the complete lifecycle and IRR. These are the model P_17-1100-WZV from Lautner Energiespartechnik GmbH and the model ROV UNI/3000-T-1,4 from KASTT, spol. s r.o with empty tube velocities from the fresh air of 2.0 m/s, which is the middle operating point from the Eurovent database. These models are typical rotors from the Eurovent Database and were successfully simulated within the SEROW Project and therefore here compared.

11.3 Results and Discussion

11.3.1 Economical Assessment

The results in this section is be normalized to the manufacturing costs for protecting sensitive company data and is done to the P_17-1100-WZV rotor from Lautner according to the general consideration on Sect. 11.2.2 and economical assessment on Sect. 11.2.3. The economical assessment results show that net savings are 4 times the investment costs. In this case the device is an attractive investment since the IRR of this rotor is 50% (Fig. 11.5).

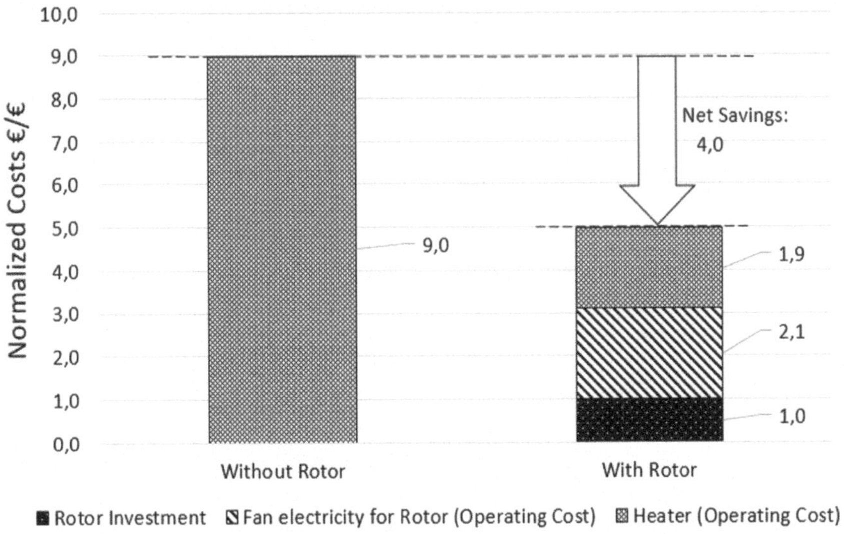

Fig. 11.5 Normalized comparison of investment and operating costs for the P_17-1100-WZV rotary heat exchanger and no use of a heat exchanger. The results in this graph are normalized to the manufacturing costs for protecting sensitive company data

11.3.2 Life Cycle Assessment

The results in this section are normalized to the manufacturing emissions for protecting sensitive company data. The results show that for the Lautner rotor and given conditions (Sects. 11.2.2 and 11.2.4 in this work), the emissions from the use phase are the most significant in the life cycle (Fig. 11.6).

Although emissions occur in the recycling phase during the transportation, metal recycling and the thermal reuse of the plastics, the recycling phase is a GWP net saving phase. The reason for this is that the recycling produces a metal substitution from primary resources that save emissions and this savings outnumber the emissions from the recycling process itself. This is shown by the −0.8 CO_2-Eq/CO_2-Eq during the recycle phase. For the transportation, 100 km distance per freight lorry is assumed between the customer and the recycling facility.

11.3.3 Optimization

Table 11.8 shows the results from both optimizations, with different objective functions (minimize GWP and maximize IRR) and a comparison to the two commercial products from the Eurovent database.

Only in terms of inclination angle and material thickness, the optimization results reach similar results: This are a high inclination angle and a low material thickness.

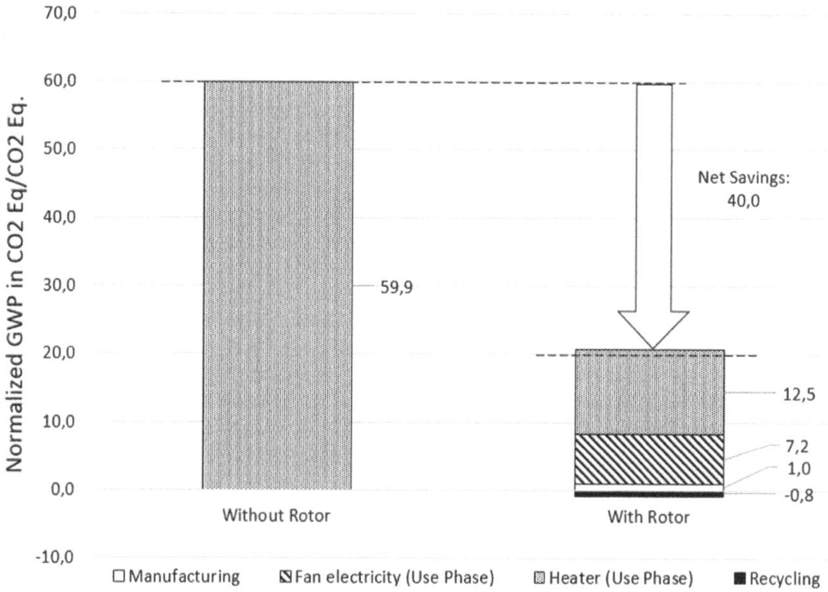

Fig. 11.6 Comparison of normalized GWP 100 emissions in the lifecycle of the P_17-1100-WZV rotary heat exchanger and no use of a rotor. The results in this graph are normalized to the manufacturing emissions for protecting sensitive company data

Table 11.8 Optimization results

Objective function	GWP	IRR	P_17	ROV
Rotor length in cm	94	15	20	20
Wave height in mm	3.9	2.8	1.4	1.4
Empty tube gas velocity ambient air in m/s	1	4	2	2
Inclination angle in °	60	60	41	46
Material thickness in μm	60	60	60	70
Savings in GWP 100 in t CO_2-eq	159	107	136	134
Internal rate of return in %	5	71	46	40

It is to notice that the rotor optimized for GWP savings has a low IRR. In contrast, the IRR optimized rotor has low GWP savings. In the IRR optimized model, the GWP savings are even lower than the current technologies (the P_17-1100-WZV and ROV UNI/3000-T-1,5 models). In the two extreme cases one can see that there is a conflict in this optimization, dependent on the defined objective function. Furthermore, if the iterations during the optimization process are graphically presented, Fig. 11.7 is obtained.

All the iterations found in Fig. 11.7 are possible configurations for rotors within the range specified previously. All these possibilities are within the range of what is

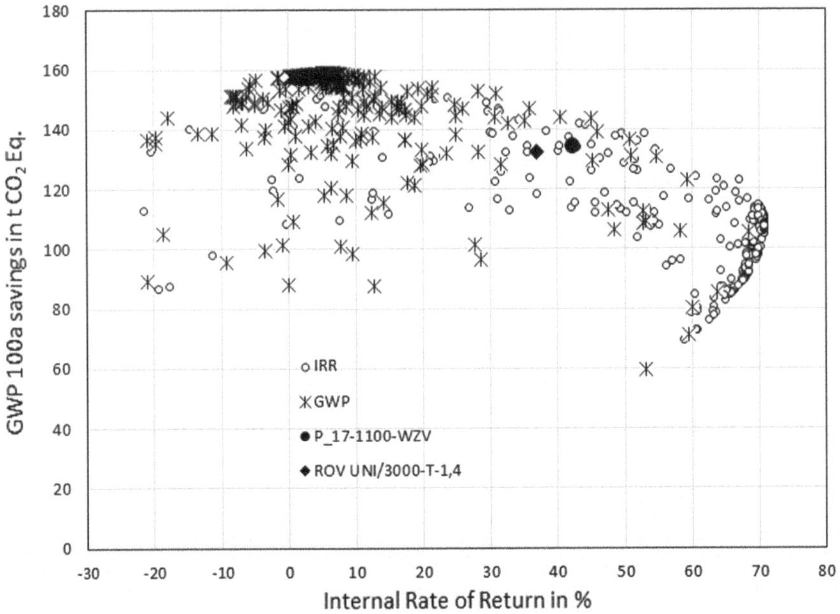

Fig. 11.7 Results of optimizations with GWP 100 and IRR as objective functions compared to the commercial rotors ROV UNI/3000-T-1,4 from Kastt and P_17-1100-WZV from Lautner

possible to manufacture, and yet a clear tendency can be shown between economic and environmental impacts. For further improvements into the optimization, a possible solution could be a multi-objective optimization approach, but this is out of the scope of this work. An interesting observation on Fig. 11.7 is that the upper points on the plotted area are other possible “solutions” for the optimization problem by doing a compromise between economic and environmental aspects. The rest of the lower points represent configurations and working conditions that are suboptimal.

In the positive side of the IRR the tendency is, the higher the IRR, the lower the GWP 100 savings. The reason behind this is that for the same volumetric flow at high gas velocities rotors with a smaller diameter can be designed. Smaller rotors require less material and are cheaper on the investment costs.

In the case of the rotor P_17-1100-WZV the optimization software shows other configurations with similar IRR but better GWP savings. In these cases the rotors could be further improved without losing economic benefits. Also an improvement for the P_17-1100-WZV could be done on the economical aspect without compromising the environmental side. Additionally, the ROV UNI/3000-T-1,4 shows similar GWP 100 savings as the P_17-1100-WZV, but the investment needed is lower on the P_17-1100-WZV. Therefore the P_17-1100-WZV model could be more economical to produce.

11.4 Conclusions

- The studied commercially available rotary heat exchangers are economically viable in a lifetime of 10 years.
- The main environmental impacts occur during the use phase.
- There is a conflict between economical optimization and environmental optimization.
- Some commercially available rotary heat exchangers could be environmentally improved without reducing the internal rate of return.
- A multi-objective optimization method could be used to improve optimization results.

Acknowledgements We would like to thank our industrial partner R. Scheuchl GmbH for their support and constructive discussions. Further on we would like to thank the Pforzheim University of Applied Sciences for their support in our laboratory and any kind of logistics. This work has been developed within the SEROW project (Reference number: 13FH049PX4). It was sponsored by the German Federal Ministry of Education and Research (BMBF) within the research program FHprofUnt.

References

Anondi GmbH Was kostet eine Kilowattstunde Gas? (2019) https://www.kwh-preis.de/gas/ratgeber/was-kostet-eine-kilowattstunde-gas. Accessed 07 Oct 2019

Dubreuil A, Young SB, Atherton J et al (2010) Metals recycling maps and allocation procedures in life cycle assessment. Int J Life Cycle Assess 15(6):621–634. https://doi.org/10.1007/s11367-010-0174-5

ecoinvent Association ecoinvent Version 3.5 (2018) Database

Eurovent Certita Certification Certified products. https://www.eurovent-certification.com. Accessed 01 Oct 2019

GET Sol 1 GmbH Strompreis pro kWh 2019. https://strom.preisvergleich.de/info/121/strompreis-kwh/. Accessed 07 Oct 2019

Graedel TE, Allwood J, Birat J-P et al (2011) What do we know about metal recycling rates? J Ind Ecol 15(3):355–366. https://doi.org/10.1111/j.1530-9290.2011.00342.x

Graedel TE, Harper EM, Nassar NT et al (2015) Criticality of metals and metalloids. Proc Natl Acad Sci U S A 112(14):4257–4262. https://doi.org/10.1073/pnas.1500415112

Hochschule Pforzheim (2019) FHProfUnt 2015 Schlussbericht: Simultane Energie- und Ressourceneffizienzoptimierung von Wärmeübertragungsregeneratoren: SEROW

O'Connor D, Calautit JKS, Hughes BR (2016) A review of heat recovery technology for passive ventilation applications. Renew Sustain Energy Rev 54:1481–1493. https://doi.org/10.1016/j.rser.2015.10.039

OPTTEK SYSTEMS I Optquest. https://www.opttek.com/products/optquest/

Perry RH, Green DW, Maloney JO (eds) (1997) Perry's chemical engineers' handbook, 7th ed. McGraw-Hill Chemical Engineering Serie Monographs, vol 23. McGraw-Hill, New York

Pufal M (2017) Rechenmodell zur ökobilanziellen und wirtschaftlichen Betrachtung eines Regenerativwärmeübertragers: Bachelorthesis

Transfactor Technologies SRL (2019) Aluminium Schrott. https://www.schrottpreis.org/aluminium/. Accessed 07 Oct 2019
VDI-Fachbereich Technische Gebäudeausrüstung (1997) Wärmerückgewinnung in Raumlufttechnischen Anlagen (VDI 2071)

Chapter 12
Pros and Cons of Batteries in Green Energy Supply of Residential Districts — A Life Cycle Analysis

Steffen Lewerenz

Abstract For the mitigation of climate change a switch to renewable energies in combination with battery storage and high efficient technologies, such as combined heat and power is necessary. To assess the environmental impacts of an electricity system model of a residential district, a Life Cycle Assessment is conducted. Different approaches of impact assessment methods are applied and compared. Based on the assumptions made, the use of a battery storage cannot always be recommended due to its dependence on the expected lifetime and capacity utilization in general. When the full cycle life of a battery storage is reached and the consumed electricity originates from photovoltaic an environmental advantage compared to the German electricity mix is created. Consequently, battery storage application can reduce the climate change potential in the conducted analysis up to −26%. A disadvantage is the introduction of a higher resource consumption compared to the German electricity mix. Noticeable is only a slight decrease in the fossil energy demand due to the utilization of natural gas by the combined heat and power plant. Furthermore, there is a displacement of mining or extraction of fossil fuels towards Russia, USA and Rest of World, which may influence the security of supply.

Keywords Battery storage · Life cycle assessment · Residential energy system models

12.1 Introduction

Climate change is one of the best-known and much-discussed cases of human impacts on the environment, which is likely to cross the threshold of reversibility soon (Steffen et al. 2015). Electricity and heat generation cause a large proportion of carbon dioxide

S. Lewerenz (✉)
Pforzheim University of Applied Sciences, 75175 Pforzheim, Germany
e-mail: steffen.lewerenz@hs-pforzheim.de

Springer Heidelberg, Tiergartenstr. 17, 69121 Heidelberg, Germany

S. Albrecht et al. (eds.), *Progress in Life Cycle Assessment 2019*,
Sustainable Production, Life Cycle Engineering and Management,
https://doi.org/10.1007/978-3-030-50519-6_12

emissions worldwide (International Energy Agency 2018). For example, in the European Union, households account for 29% of total electricity consumption (eurostat 2017), which is still generated to 74% by conventional power plants (eurostat 2019a).

The substitution of conventional power plants by renewable energies includes the use of energy sources, such as solar and wind energy. In order to overcome their fluctuations and the temporal shift between electricity generation and consumption, electricity storage systems are needed (Samsatli and Samsatli 2018). A large number of storage technologies are available that cause environmental impacts through their production and use, such as resource consumption and climate change (Baumann et al. 2017; Weber et al. 2018; Spanos et al. 2015).

Furthermore, combined heat and power plants have the potential to reduce climate change even though they are fired by fossil fuels. Especially for the heat generation in residential districts, it is already applied and for the German ministry of economics and energy it is considered as a technology bridging into a less carbon dioxide extensive future (Bundesministerium für Wirtschaft und Energie 2019).

Life Cycle Assessment is a commonly applied method to environmentally assess energy systems (Norm DIN EN ISO 14044 2018; Norm DIN EN 14040 2019). Amongst others Rauner and Budzinski (2017), Berrill et al. (2016) and Volkart et al. (2017) apply Life Cycle Assessment for the analysis of national (Rauner and Budzinski 2017; Volkart et al. 2017; Vandepaer et al. 2018), European (Berrill et al. 2016), and international (Volkart et al. 2018) energy systems. Different approaches of impact assessment methods are used: varying from the use of several midpoint indicators considering the environmental impact (Rauner and Budzinski 2017), over the integration of social and security of supply indicators (Volkart et al. 2017) up to the application of single score indicators (Rauner and Budzinski 2017).

As Life Cycle Assessments for residential districts, including the use of battery storage, are rarely carried out, the question arises whether or not integration has a positive impact on the environmental impact of an energy system. Lithium-ion-iron-phosphate and vanadium-redox-flow battery are two appropriate electricity storage systems, which are already in use for stationary applications (McManus 2012; Cunha et al. 2015).

This paper processes results of the research project "urban energy systems and resource efficiency" (ENsource), which has the goal to develop a service-oriented model architecture to solve coupled distributed simulation, optimization and energy management in combination with the assessment of impacts on the environment. The ENsource team consists of a cross-university research network, whereby the Pforzheim University focusses on the environmental assessment of energy systems. Thus, this paper environmentally assesses the integration of the above-mentioned battery storage into a residential energy system based on dispatch optimization results. Furthermore, by investigating the displacement effects of the utilized energy carriers, especially fossil fuels, the security of supply aspect is briefly mentioned.

12.2 Method

The applied method is a combination of energy system optimization and Life Cycle Assessment (Norm DIN EN 14040 2019). The energy system model consists of 74 households, photovoltaic plants, a combined heat and power plant, the electricity grid and different electricity storage systems: a lithium-ion-iron-phosphate battery (LFP) and a vanadium-redox-flow battery (VRFB). The dispatch of the energy system model is optimized based on variable costs applying the "Open Energy Modelling Framework", which was developed at the University of Applied Sciences in Flensburg (Hilpert et al. 2018). Needed specific economic and technologic parameter are based on a literature research. All parameters and assumptions are displayed in Fig. 12.1. For a full description of the designed optimization model with all parameters and assumptions refer to Lewerenz (Lewerenz 2019). The underlying optimization model is parameterizable and can be calculated with any other available current load profiles.

12.2.1 *Life Cycle Assessment*

To quantify potential environmental impacts, Life Cycle Assessment is a frequently applied approach (Weber et al. 2018; Majeau-Bettez et al. 2011; Peters et al. 2017). As all stages of life from procurement, production, use and disposal of a product or process are assessed, a holistic assessment is conducted. All inputs including material

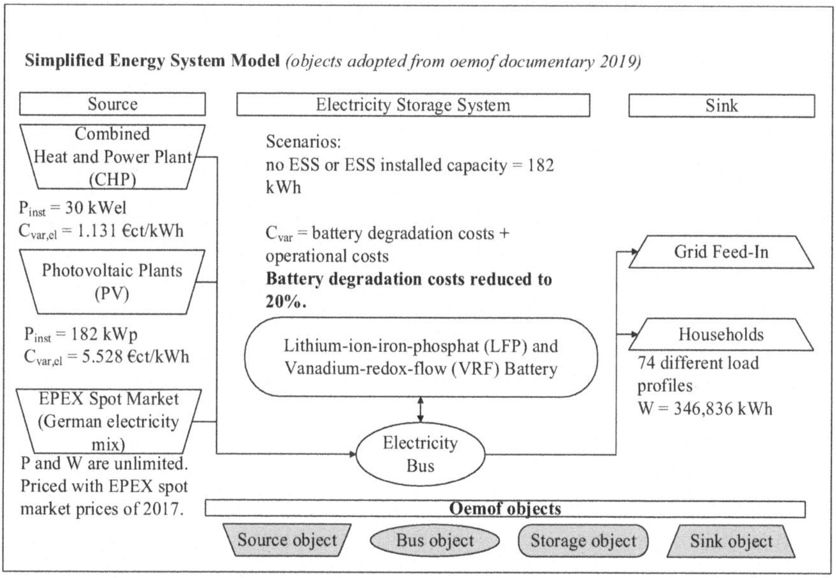

Fig. 12.1 Parameters and assumptions (Lewerenz 2019)

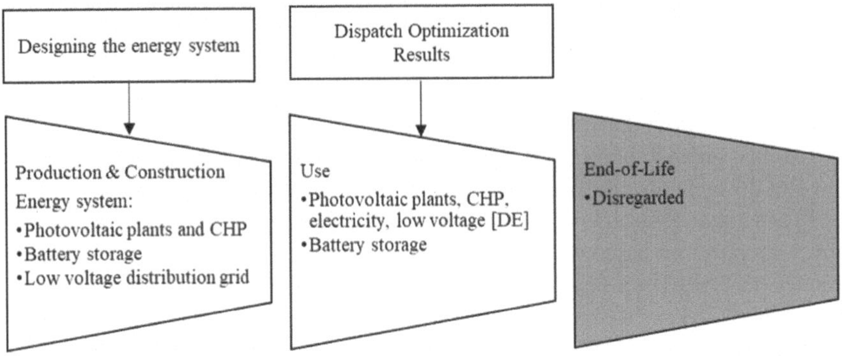

Fig. 12.2 Analyzed product system

and energy must be taken into account (Norm DIN EN ISO 14044 2018). The Life Cycle Assessment is based on ISO 14044, which provides a framework that describes how to carry out a Life Cycle Assessment (Norm DIN EN 14040 2019; Zackrisson et al. 2010).

The goal of the assessment is to provide an analysis of different electricity systems to supply a residential district. The analyzed product system with its boundaries is depicted in Fig. 12.2. The production and use phase of the described product system are analyzed. The end-of-life stage is disregarded, as there are either no reliable sources for the analyzed battery storage available or only limited information in literature (Baumann et al. 2017; Weber et al. 2018). The designed power system with all its parameters is the basis for the production and construction phase of the Life Cycle Assessment. All dispatch optimization results for example the total electricity generated by the power plants including the charge into the battery storage system are inputs for the use phase of the product system.

The functional unit is defined as "The complete power supply of the defined residential district over one year".

Modelling. To model the production of the electricity generation and distribution data from ecoinvent 3.3 is utilized (Wernet et al. 2016). The installed power of the power generation plants and the capacity of the battery storage is fixed; thus the production is independent from the use phase. The use phase for each scenario is modelled according to the obtained dispatch results. Amongst others, natural gas consumption, lubricating oil of the combined heat and power plant as well as the water demand of the photovoltaic plant is considered—in line with the ecoinvent data sets.

The 182 kWp photovoltaic plants are modeled according to the ecoinvent process: "market for photovoltaic slanted-roof installation, 3 kWp, multi-Si, panel, mounted, on roof, cut-off, U—GLO". The lifetime of the power plant is set to 30 years (Jungbluth et al. 2019). The annual yield of the photovoltaic plant is calculated with the PV GIS (European Commission 2006). Overproduced photovoltaic electricity, which is fed into the electricity grid receives no credit. The combined heat and power plant is modelled according to the global ecoinvent processes for a 50 kW combined heat and power unit including common components for heat and electricity, components for electricity and components for heat only. It is linearly scaled to 30 kW, which is the installed power of the combined heat and power plant used in the electricity system model. The lifetime accounts for 15 years (Norm VDI 2067 2012). The distance of the distribution network (low voltage) is assumed to have the same length as in the market process of the German electricity mix.

Battery storage modelling. The lifetime of the battery storage is modelled with two different approaches. The first approach is based on the calendric lifetime of the battery, which is assumed to be 10 years (Lewerenz 2019). The second approach is using the watt-hours throughput model (Wh-model), which considers the cycle life of batteries (Bindner 2005; Bordin et al. 2017). The Wh-model is a simple battery model, which allows the calculation of the total amount of (dis)charged watt-hours through the battery storage (Spanos et al. 2015). In addition, it is not necessary to provide exact information about the battery storage, as it neglects the charging and discharging processes (de Beer and Rix 2016). By utilizing the lifetime curve of the battery the amount of cycles to failure depending on different depth of discharge until the end of life of the battery can be calculated. The data is obtained from the datasheets of battery storage (Bindner 2005). The datasheets for the battery storage Voltstorage SMART (Technisches Datenblatt 2019) and the TR12.6-92 (Trojan Battery Company 2019) are used, which are representing a VRFB and a LFP respectively.

The total amount of possible watt-hour throughput was calculated at the design phase of the electricity system (Lewerenz 2019). As a result of the dispatch optimization the amount of used watt-hour throughput is determined. Consequently, by dividing the used watt-hours by the total possible amount of watt-hour results in the total lifetime in years for the battery storage. That lifetime accounts 24 years for the LFP-system and 42 years for the VRFB-system.

The life cycle inventory for the battery storage is essentially grounded on a literature research and the ecoinvent 3.3 database (Wernet et al. 2016). To model the LFP-system the inventory published from (Peters and Weil 2018) is used. They developed unified inventories for lithium-ion-batteries by analyzing different studies of LFP batteries (Peters and Weil 2018). For example, (Zackrisson et al. 2010) describe the manufacturing process for 1 kg of LFP, which is used to model the applied LFP TR 12.8 V/92Ah (Zackrisson et al. 2010). The VRFB-system is modelled according to Weber et al. (2018) utilizing their published inventory for the production of 1 kg of VRFB.

A 600 km transport distance from the production place to the end users is assumed (eurostat 2019b). Processes delivering energy are modelled when possible with European energy mixes as the production is assumed to take place in Europe. In the background system preferably market processes are applied, which consist of a global supply chain (Wernet et al. 2016).

Scenarios. Five scenarios are analyzed. The first scenario is the designed electricity system model without the application of a battery storage (Base). The second and third scenario are utilizing the LFP-system with an assumed lifetime of 10 years (LFP (10a)) and based on the cycle life (LFP (cycles)). The fourth and fifth scenario is

Table 12.1 Input parameters for use phase (Lewerenz 2019)

Electricity generation in MWh	Reference	Base	LFP (10a)	LFP (cycles)	VRFB (10a)	VRFB (cycles)
CHP (direct use)	–	160	160	160	160	160
CHP (charge)	–	–	5	5	10	10
PV (direct use)	–	102	102	102	102	102
PV (feed-in)	–	79	79	57	50	50
PV (charge)	–	–	23	23	30	30
German electricity mix	347	85	62	62	66	66

CHP—combined heat and power; PV—photovoltaic plant

operating the VRFB-system also with an assumed lifetime of 10 years (VRFB (10a)) and based on the cycle life (VRFB (cycles)). The reference scenario is the full electricity supply with the German electricity mix. For each scenario input parameters for the use phase are displayed in Table 12.1.

Impact Assessment Method. In the ENsource project a comprehensive set of indicators was generated. This set consists of impact indicators, which are suitable to evaluate an energy system (Hottenroth and Lambrecht 2016; Zentrum für angewandte Forschung Urbane ENergiesysteme und Ressourceneffizienz - ENsource 2019). The indicators can be clustered in five main categories: energy demand, resource consumption, water consumption, land use and ecosystem service. A full list of the ENsource indicator set is displayed in Table 12.2. For the assessment of the analyzed energy system the cumulated energy demand is substituted by the non-renewable energy demand including fossil, nuclear and primary forest as the analysis is focused on the non-renewable energy.

For further analysis, the deviations of all scenarios from the reference scenario (German electricity mix) are calculated. The arithmetic mean of the deviations is then determined. This is done for each impact category, which is composed of the corresponding impact indicators. Finally, the arithmetic mean of all impact indicators is determined and compared with the impact assessment method ecological scarcity 2013 (Büsser-Knöpfel et al. 2013; Frischknecht et al. 2006).

Table 12.2 Applied impact assessment categories and indicators (Hottenroth and Lambrecht 2016)

Impact category	Impact indicator	Impact assessment method
Energy demand (non-renewable)	Fossil energy demand	VDI 4600 (Norm VDI 4600 2012)
	Nuclear energy demand	VDI 4600 (Norm VDI 4600 2012)
	Primary forest energy demand	VDI 4600 (Norm VDI 4600 2012)
Resource consumption	Cumulated resource consumption	VDI 4800 (Norm VDI 4800 2016)
	Abiotic depletion potential	ILCD (International Reference Life Cycle Data System (ILCD) 2011)
Water consumption	Aggregated inventory result	
Land	Land use	ILCD (International Reference Life Cycle Data System (ILCD) 2011)
	Land competition	CML 2001 (Guinée 2002)
Ecosystem service	Climate change (GWP 100a)	ILCD (International Reference Life Cycle Data System (ILCD) 2011)
	Freshwater and terrestrial acidification	ILCD (International Reference Life Cycle Data System (ILCD) 2011)
	Freshwater ecotoxicity	ILCD (International Reference Life Cycle Data System (ILCD) 2011)
	Freshwater eutrophication	ILCD (International Reference Life Cycle Data System (ILCD) 2011)
	Ozone Layer depletion	ILCD (International Reference Life Cycle Data System (ILCD) 2011)
	Photochemical ozone creation	ILCD (International Reference Life Cycle Data System (ILCD) 2011)
	Respiratory effects, inorganics	ILCD (International Reference Life Cycle Data System (ILCD) 2011)

12.3 Results

12.3.1 Life Cycle Assessment

The deviations of each impact category compared to the reference scenario are displayed in Fig. 12.3. The approach used to calculate the arithmetic mean using the impact indicators leads to a reduction in the overall impact for all scenarios: Base (−17%), VRFB(10a) (−9%), LFP (10a) (−8%), VRFB(cycles) (−18%) and LFP(cycles) (−17%). The applied impact assessment method ecological scarcity 2013 results in higher reductions from −23% (Base) to −28% (VRFB/LFP (cycles)). Consequently, the approach of the arithmetic mean indicates, that the storage utilization is not environmentally beneficial compared to the Base scenario, when its lifetime is 10 years. It is environmentally viable when the full amount of cycles can be deployed. But there is no further decrease of environmental impacts generated

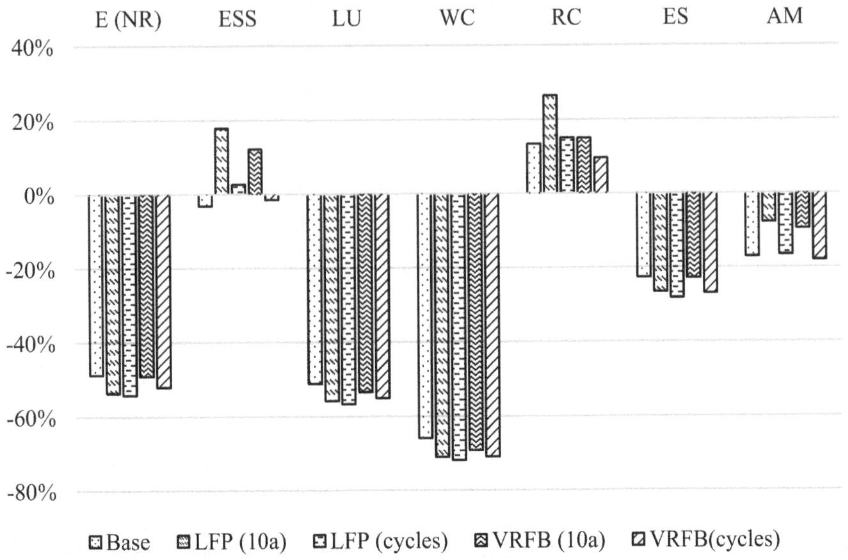

Fig. 12.3 Impact results for each impact category for all scenarios. (E (NR) = Energy (non-renewable); ESS = ecosystem service; LU = land use; WC = water consumption; RC = resource consumption; ES = ecological scarcity 2013; AM = arithmetic mean)

by their utilization, thus an application of a battery storage is not recommended. In contrast, the ecological scarcity method indicates that the utilization of the battery storage is preferable for all scenarios compared to the German electricity mix. Reductions of impacts, compared to the Base scenario, in the scenarios LFP(10a) (−4%), LFP(cycles) (−5%) and VRFB(cycles) (−4%) are gained.

Impacts on water consumption, land use, global warming potential, and non-renewable energy demand are reduced over all scenarios. The resource consumption is increased over all scenarios compared to the reference scenario. Concerning the ecosystem service, the Base and VRFB(cycles) can slightly reduce the impact whereas the other scenarios worsen the results.

In general, the environmental impact of the power system depends on the lifetime of the battery storage when used. If the VRFB and the LFP battery storage can reach their full cycle lifetime, the overall impact of the VRFB can be reduced, while the LFP battery storage has the same environmental impact compared to the Base scenario (using arithmetic mean). The shorter battery life leads to a deterioration of the environmental performance compared to the Base scenario. Compared to these results, the ecological scarcity method leads to a better (or the same for VRFB(10a)) environmental performance for each battery used. When setting the battery life to 10 years, it is noticeable that the LFP battery is able to reduce the environmental impact by 4% compared to the Base scenario. The application of the VRFB delivers the same overall environmental impact compared to the Base scenario.

Resource Consumption. The indicator abiotic depletion potential shows an increased zinc-lead mine operation and tantalum production to produce the photovoltaic plants as well as the inverter in the Base scenario. The utilization of battery storage even further increase the resource consumption by 2% (VRFB(10a)) and 13% (LFP(10a)). Only in the scenario VRFB(cycles) is a slight decrease in resource consumption of 2% noticeable. Contrary to the abiotic depletion potential, the cumulated resource consumption is exhibiting the highest impacts due to tantalum, which is mainly used in the inverter production. The dominant process is the capacitor production. Copper mine ferronickel production is next to tantalum the strongest emission contributing to that indicator. In contrast to the abiotic depletion potential the cumulated resource consumption is not driven by zinc-lead mine operation. When utilizing the VRFB the resource consumption is further increased by the vanadium bearing magnetite production.

Water Consumption. The water consumption decreases by up to 72% in the scenario LFP(cycles) and 71% (VRFB(cycles)) respectively. This is mainly due to the substitution of the German electricity mix. Consequently, less decarbonized water is needed for the electricity generation in lignite, hard coal and nuclear power plants.

Land Use. In total the land use decreases due to a high reduction of utilized biomass, explicitly hardwood, softwood forestry and rape seed production. However, as a result of an increased power generation with natural gas the indicator land use by oil and natural gas production increases because of a higher amount of on/offshore oil and gas wells. Furthermore, the land use reduces due to the decline of photovoltaic open ground installations, which are implemented in the market dataset for the German electricity mix.

The same result is obtained for the CML indicator land competition with even higher effects, which lead to a possible reduction of −57% for the scenario LFP(cycles).

Ecosystem Service. The analysis of the impact category ecosystem service displays a diverse picture. In the scenarios Base and VRFB(cycles) the impact reduces by −1% and −3% respectively. All other scenarios exhibit a higher impact varying from 3% (LFP(cycles)) to 18% (LFP(10a)). The assessed impact categories, freshwater and terrestrial acidification, ozone layer depletion, photochemical ozone creation and respiratory effects are increasing for all scenarios compared to the reference scenario. The growth is mainly due to a higher natural gas production and the production of the photovoltaic system. Noticeable is a high increase of the indicator ozone layer depletion especially when applying the LFP-system with a lifetime of 10 years (160%). In the scenarios Base and VRFB(cycles) the increase is much lower and accounts for 5% and 19% respectively.

In the scenarios LFP(10a) and LFP(cycles) the impact category climate change reduces up to −26%. The Base and the VRFB(10a) scenario exhibit a reduction of −22%, whereas the scenario VRFB(cycles) can reach a reduction of −24%. These reductions are based on the substitution of electricity from lignite and hard coal with electricity from natural gas and photovoltaics. Furthermore, reductions in the impact categories freshwater ecotoxicity and freshwater eutrophication are obtained, which account for −43% (Base) to −49% (LFP(cycles)) and −70% (Base) to −76%

(LFP(cycles)) respectively. Those reductions are mainly due to the reduction of the utilization of lignite for electricity generation.

Energy (non-renewable). The assessment results in a decrease of utilized non-renewable energy carriers. The lowest reduction is reached in the Base scenario with −49%, the highest reduction in the scenarios VRFB(cycles), LFP(cycles) and LFP(10a) and accounts for −54%. Especially energy from nuclear (average: −23%) and primary forest (average: −27%) is reduced noticeably due to the substitution of the German electricity mix.

When analyzing the used non-renewable fossil energy only a slight reduction of −4% (Base) to −8% (LFP/VRFB(cycles)) is obtained. As displayed in Fig. 12.4 a shift from lignite and hard coal towards natural gas can be noticed when comparing the German electricity mix and the Base scenario. Almost all energy generated by hard coal and lignite is thus substituted by natural gas. As a consequence, also a geographic displacement of the origin of energy carriers is noticeable: hard coal and lignite, which is mainly mined in Western Europe and Europe, is substituted by natural gas from Russia, USA, Rest of World and Global. Furthermore, an increased use of Chinese hard coal is established, which is due to an increased production of photovoltaic systems in China.

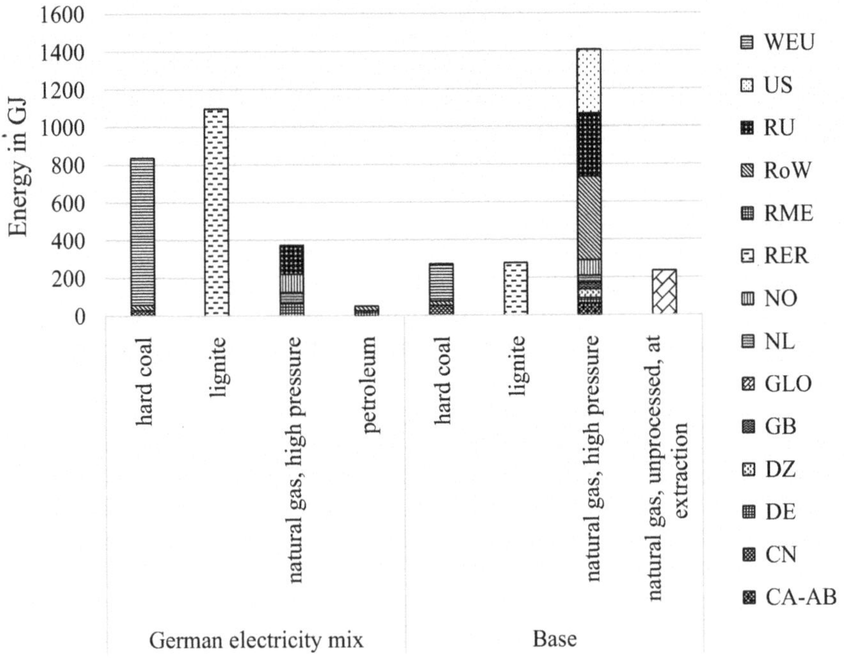

Fig. 12.4 Results for non-renewable fossil energy for the main processes and countries. (WEU = Western Europe; US = USA; RU = Russia; RoW = Rest of World; RME = Middle East; RER = Europe; NO = Norway; NL = Netherlands; GLO = Global; GB = Great Britain; DZ = Algeria; DE = Germany; CN = China; CA-AB = Canada-Alberta)

12.4 Discussion

Overall it can be concluded, that a switch to electricity generated by a combination of combined heat and power and photovoltaic is beneficial in comparison with the German electricity mix. There is no explicit statement or conclusion, which argue for or against the utilization of battery storage. The distance to target method ecological scarcity indicates that the utilization of the battery storage with both investigated lifetimes should be applied. The ecological scarcity method always recommends the utilization of battery storage, which results from the lower weights of the impact indicators like ozone layer depletion or abiotic resource depletion. Especially, high impacts are obtained for the impact indicator ozone layer depletion. This is based on the binder production for the electrode of the LFP system and results from the market process of tetrafluoroethylene, which represents the substitution of the actually used substance polyvinylidene fluoride and is subject to high uncertainties (Peters and Weil 2018).

Additionally, a higher resource consumption is obtained, when integrating a battery storage to the more resource intensive renewable energy power plants. The increase in abiotic resource depletion is mainly based on the utilization of photovoltaic, which leads to a higher zinc, copper and tantalum mine operation.

The analysis of the non-renewable energy carriers exhibits a distinct decrease in the energetical use of nuclear energy and primary forest. Although, the consumption of fossil fuels is only slightly decreasing. As a result of the installed natural gas fired combined heat and power plant, a high gas consumption is introduced. This leads to a clear switch from lignite and hard coal towards natural gas, which is mainly extracted in the USA, Russia and Rest of World. Subsequently, the "German coal commission" demanded switch to natural gas can introduce security of supply problems and a burden shifting towards developing countries, which might be included in Rest of World. Finally, an increased utilization of Chinese hard coal implies another shift of energy generation. Subsequently, by the described movement the dependency on other states is noticeably growing, although the total fossil energy demand is almost not decreasing.

If the arithmetic mean is calculated over all indicators, no significant advantage can be seen from the use of battery storage compared to the Base scenario. Nevertheless, there is a dependency on life expectancy, as shorter life expectancies (LFP/VRFB(10a)) produce a higher environmental impact than longer life expectancies (LFP/VRFB(cycles). Consequently, the utilization of battery storage is environmentally viable compared to the German electricity mix as long as it can utilize all of its possible cycles and the German electricity mix is substituted. Although the LFP(10a) is able to reduce the environmental impacts slightly (4%) compared to the Base scenario, the environmental sustainability is still questionable, as the reduction is small and, in addition, the end-of-life processes are neglected. If the battery system reaches the end of its service life after 10 years, e.g. because the calendrical service life ends, an environmentally sound application is uncertain. Consequently, when implementing a battery storage in an energy system it should be designed to reach its possible cycling lifetime. The question on, what the suitable expected lifetime to model a battery storage is, must be further researched. Furthermore, the end of life of a battery is reached, when only 80% of its initial capacity is still available (Reid and Julve 2016; Bobba et al. 2018). Especially, when utilizing discarded traction batteries of electric vehicles as stationary battery systems (Bobba et al. 2018), their lifetime is expanded and 80% might not be a sufficient assumption anymore.

Furthermore, there is a trade-off between the economic and ecological advantages of battery storage. If the electricity from the storage is generated by a renewable energy source and replaces the German electricity mix, an ecological advantage can be observed. However, the utilization might not be economically viable, when the electricity from the grid is less cost-intensive.

12.5 Conclusion and Outlook

Two different aggregation approaches for impact indicators were analyzed and compared, resulting in different conclusions concerning the application of battery storage. Further research is needed to assess the potential environmental impacts of renewable energy systems and battery storage and to provide clearer answers. Especially the increasing resource demand must be assessed in more detail. For instance, resource criticality, geopolitical (e.g. country allocation of resources), political factors (e.g. political stability) (Norm VDI 2067 2012) and social factors (e.g. social stability) (Volkart et al. 2017) could be integrated.

At present the applied impact assessment method ecological scarcity is based on environmental political targets of Switzerland. The method could be adopted for German energy system, for instance by applying and adapting the method of ecological scarcity for Germany developed by Ahbe (2014), Ahbe et al. (2018).

References

Ahbe S (2014) Methode der ökologischen Knappheit für Deutschland. Umweltbewertungen in Unternehmen; eine Initiative der Volkswagen AG. 2., überarbeitete Auflage. Logos-Verlag, Berlin

Ahbe S, Weihofen S, Wellge S (2018) The ecological scarcity method for the European Union. Springer Fachmedien Wiesbaden, Wiesbaden

Baumann M, Peters JF, Weil M, Grunwald A (2017) CO_2 footprint and life-cycle costs of electrochemical energy storage for stationary grid applications. Energy Technol 5(7):1071–1083

Berrill P, Arvesen A, Scholz Y, Gils HC, Hertwich EG (2016) Environmental impacts of high penetration renewable energy scenarios for Europe. Environ Res Lett 11(1):14012

Bindner H (2005) Lifetime modelling of lead acid batteries. Risø National Laboratory, Roskilde

Bobba S, Mathieux F, Ardente F, Blengini GA, Cusenza MA, Podias A, Pfrang A (2018) Life cycle assessment of repurposed electric vehicle batteries: an adapted method based on modelling energy flows. J Energy Storage 19:213–225

Bordin C, Anuta HO, Crossland A, Gutierrez IL, Dent CJ, Vigo D (2017) A linear programming approach for battery degradation analysis and optimization in offgrid power systems with solar energy integration. Renew Energy 101:417–430

Bundesministerium für Wirtschaft und Energie: Kommission „Wachstum, Strukturwandel und Beschäftigung". Abschlussbericht, https://www.bmwi.de/Redaktion/DE/Downloads/A/abschlussbericht-kommission-wachstum-strukturwandel-und-beschaeftigung.pdf?__blob=publicationFile, last accessed 20.05.2019

Cunha Á, Martins J, Rodrigues N, Brito FP (2015) Vanadium redox flow batteries: a technology review. Int J Energy Res 39(7):889–918

de Beer B, Rix A (2016) Influences of energy throughput on the life of various battery technologies

European Commission, Joint Research Centre, Renewable Energies Unit: energy rating of PV modules based on PVGIS irradiance and temperature database. 21st European Photovoltaic Solar Energy Conference, pp 2088–2092 (2006)

eurostat (2017) Energy balance sheets. 2015 DATA. 2017 edn. Publications Office of the European Union, Luxembourg

eurostat (2019a) Electricity generation statistics. European Union, https://ec.europa.eu/eurostat/statistics-explained/images/7/73/Electricity_Statistics_Files%2C_2017_%28GWh%29-1-update-23-04-2018.xlsx, last accessed 14.03.2019

eurostat (2019b) Road freight transport by journey characteristics. https://ec.europa.eu/eurostat/statistics-explained/index.php/Road_freight_transport_by_journey_characteristics#Average_distance_travelled, last accessed 16.04.2019

Frischknecht R, Büsser-Knöpfel, Sybille, treeze Ltd.: Swiss Eco-Factors 2013 according to the Ecological Scarcity Method, Bern, https://www.bafu.admin.ch/bafu/en/home/topics/economy-consumption/economy-and-consumption–publications/publications-economy-and-consumption/eco-factors-2015-scarcity.html, last accessed 10.10.2019

Frischknecht R, Steiner R, Jungbluth N (2006) The ecological scarcity method—eco-factors 2006. A method for impact assessment in LCA. Environ Stud No 906 (2009)

Guinée JB (2002) Handbook on life cycle assessment. Operational guide to the ISO standards. Kluwer Academic Publishers, Dordrecht, Boston (op. 2002)

Hilpert S, Kaldemeyer C, Krien U, Günther S, Wingenbach C, Plessmann G (2018) The open energy modelling framework (OEMOF)—a new approach to facilitate open science in energy system modelling. Energy Strategy Rev 22:16–25

Hottenroth H, Lambrecht H (2016) Meilensteindokument ENsource - M 3.1 Indikatorenset und Daten für das Gesamtprojekt und zur Anwendung in Fallstudien. HS Pforzheim

International Energy Agency (2018) CO_2 emissions from fuel combustion. Highlights, https://www.iea.org/publications/freepublications/publication/CO2EmissionsfromFuelCombustionHighlights2017.pdf, last accessed 24.11.2018

International Reference Life Cycle Data System (ILCD) (2011) handbook. General guide for life cycle assessment: provisions and action steps, 1st edn. (EUR, Scientific and technical research series, 24571). Publications Office, Luxembourg

Jungbluth N, Stucki M, Flury K, Frischknecht R, Büsser S (2019) Life cycle inventories of photovoltaics, http://esu-services.ch/fileadmin/download/publicLCI/jungbluth-2012-LCI-Photovoltaics.pdf, last accessed 10.10.2019

Lewerenz S (2019) Model based dispatch optimisation for residential districts—analysing the integration of electricity storage systems and their environmental impact. In: International Assoziation for Energy Economics (eds) Energy challenges for the next decade. Ljubljana, Slovenia, 25–28 Aug 2019. International Assoziation for Energy Economics

Majeau-Bettez G, Hawkins TR, Strømman AH (2011) Life cycle environmental assessment of lithium-ion and nickel metal hydride batteries for plug-in hybrid and battery electric vehicles. Environ Sci Technol 45(10):4548–4554

McManus MC (2012) Environmental consequences of the use of batteries in low carbon systems: the impact of battery production. Appl Energy 93:288–295

Norm VDI 2067 (2012) Wirtschaftlichkeit gebäudetechnischer Anlagen Grundlagen und Kostenberechnung, last accessed 21.12.2018
Norm DIN EN ISO 14044 (2018) Umweltmanagement - Ökobilanz-Anforderungen und Anleitungen (ISO 14044:2006 + Amd 1:2017); Deutsche Fassung EN ISO 14044:2006 + A1: 2018
Norm DIN EN ISO 14040 (2019) Umweltmanagement –Ökobilanz – Grundsätze und Rahmenbedingungen (ISO 14040:2006) Deutsche und Englische Fassung EN ISO 14040:2006
Norm VDI 4600 (2012) Kumulierter Energieaufwand (KEA)
Norm VDI 4800 (2016) Ressourceneffizienz - Methodische Grundlagen, Prinzipien und Strategien
Peters J, Weil M (2018) Providing a common base for life cycle assessments of Li -Ion batteries. J Clean Prod 171:704–713
Peters J, Baumann M, Zimmermann B, Braun J, Weil M (2017) The environmental impact of Li-Ion batteries and the role of key parameters—a review. Renew Sustain Energy Rev 67:491–506
Rauner S, Budzinski M (2017) Holistic energy system modelling combining multi-objective optimization and life cycle assessment. Environ Res Lett 12(12):1–13
Reid G, Julve J (2016) Second life-batteries as flexible storage for renewables energies. https://www.bee-ev.de/fileadmin/Publikationen/Studien/201604_Second_Life-Batterien_als_flexible_Speicher.pdf, last accessed 14.04.2019
Samsatli S, Samsatli NJ (2018) A general mixed integer linear programming model for the design and operation of integrated urban energy systems. J Clean Prod 191:458–479
Spanos C, Turney DE, Fthenakis V (2015) Life-cycle analysis of flow-assisted nickel zinc-, manganese dioxide-, and valve-regulated lead-acid batteries designed for demand-charge reduction. Renew Sustain Energy Rev 43:478–494
Steffen W, Richardson K, Rockström J, Cornell SE, Fetzer I, Bennett EM et al (2015) Sustainability. Planetary boundaries: guiding human development on a changing planet. Science 347(6223):1259855-1–1259855-10
Technisches Datenblatt. https://voltstorage.com, last accessed 09.04.2019
Trojan Battery Company, L. L.C.: TR 12.8-92 Li-ion. https://www.trojanbattery.com/pdf/datasheets/24_Lithium_12.8-92.pdf, last accessed 15.02.2019
Vandepaer L, Cloutier J, Bauer C, Amor B (2018) Integrating batteries in the future Swiss electricity supply system: a consequential environmental assessment. J Ind Ecol 18(3):321
Volkart K, Weidmann N, Bauer C, Hirschberg S (2017) Multi-criteria decision analysis of energy system transformation pathways: a case study for Switzerland. Energy Policy 106:155–168
Volkart K, Mutel CL, Panos E (2018) Integrating life cycle assessment and energy system modelling: methodology and application to the world energy scenarios. Sustain Prod Consum 16:121–133
Weber S, Peters JF, Baumann M, Weil M (2018) Life cycle assessment of a vanadium redox flow battery. Environ Sci Technol 52(18):10864–10873
Wernet G, Bauer C, Steubing B, Reinhard J, Moreno-Ruiz E, Weidema B (2016) The ecoinvent database version 3 (part I): overview and methodology. Int J Life Cycle Assess 21(9):1218–1230
Zackrisson M, Avellán L, Orlenius J (2010) Life cycle assessment of lithium-ion batteries for plug-in hybrid electric vehicles—critical issues. J Clean Prod 18(15):1519–1529
Zentrum für angewandte Forschung Urbane ENergiesysteme und Ressourceneffizienz - ENsource (2019) Abschlussbericht. Zentrum für angewandte Forschung Urbane ENergiesysteme und Ressourceneffizienz – ENsource. Hochschule für Technik Stuttgart, Stuttgart

Chapter 13
Asking Instead of Telling — Recommendations for Developing Life Cycle Assessment Within Technical R&D Projects

Miriam Lettner and Franziska Hesser

Abstract Sustainability assessment for emerging technologies is a prerequisite for sustainable process and product development. Life Cycle Assessment (LCA) is an accepted method for quantifying and assessing the potential environmental impacts of product systems. In general, an LCA is conducted on mature technologies or an industrial scale. Nevertheless, the design or development stage provides important information on potential environmental impacts and is rated to determine the grand part of environmental impacts of new technologies and products. In this study, practice-oriented recommendations for LCA practitioners and other project partners of inter-disciplinary and inter-organizational R&D projects are derived in order to enhance the integration and implementation of LCAs in technical R&D projects. The observations are presented as three main recommendations: Create common knowledge, consider LCA as part of the R&D and adapt and revise. LCA at the R&D stage can contribute to enhanced knowledge sharing and, most importantly, can support the entire R&D project by generating information on the environmental performance of the new processes and products and revealing the window of opportunity for sustainable process and product development.

Keywords Bio-based materials · Eco-design · Life cycle assessment · Lignin · R&D · Streamlined life cycle assessment · Sustainability · Technological innovation

---

M. Lettner
University of Natural Resources and Life Sciences Vienna, Institute of Marketing and Innovation, Gregor-Mendel-Straße 33, 1180 Vienna, Austria

University of Applied Sciences Kufstein Tirol, Andreas-Hofer-Straße 7, 6330 Kufstein, Austria

F. Hesser (✉)
Wood K plus — Competence Center for Wood Composites and Wood Chemistry, Kompetenzzentrum Holz GmbH, Altenberger Straße 69, 4040 Linz, Austria
e-mail: f.hesser@wood-kplus.at

S. Albrecht et al. (eds.), *Progress in Life Cycle Assessment 2019*, Sustainable Production, Life Cycle Engineering and Management, https://doi.org/10.1007/978-3-030-50519-6_13

13.1 Introduction

Over the last years, Life Cycle Assessment (LCA) has evolved from an attributional tool into a wide-ranging methodology for exploring potential environmental impacts of various processes and products (McManus and Taylor 2015). Its application is no longer only company or consumer-driven but also driven by policies across the globe (McManus and Taylor 2015; see also European Platform on LCA, European Commission 2019). In addition, the incorporation of sustainability considerations is recognized as the driving force behind the innovation of business models (Sempels and Hoffmann 2013). In the spotlight of the emerging bioeconomy, sustainability has become a requirement for competitive companies (Lacasa et al. 2016). Thus, LCA has become a valuable tool to support decision-making processes, identify potential bottlenecks and implement future-oriented thinking by identifying and assessing future improvements (Luz et al. 2018; Zafeirakopoulos and Genevois 2015). LCAs have been carried out in various sectors and industries, such as the automobiles sector (Danilecki et al. 2017; Ding et al. 2016; Schöggl et al. 2017; Zah et al. 2007), packaging (Albrecht et al. 2013; Siracusa et al. 2014; Wikström et al. 2016), agriculture and forestry (González-García et al. 2009; Karjalainen 2001; Meisterling et al. 2009; Piringer et al. 2016) or waste management (Cherubini et al. 2009; Laurent et al. 2014; Scherhaufer et al. 2018; Unger et al. 2017), etc. However, most of the LCAs focus on analyzing existing and well-established mature processes and products. LCA is used to implement eco-design for the improvement of the environmental performance of products. Shortcomings are reported with LCA in eco-design involving available knowledge and costs of decisions, expertise of the LCA practitioner, degree of novelty of the product and the conception of LCA (McAloone and Pigosso 2017). Completely new products challenge LCA as an eco-design tool due to the lack of a benchmark (previous product release, product of competitor) (Trappey et al. 2011). Arvidsson et al. (2018) recently identified the need to clarify the term "prospective LCA" for assessing emerging technologies and developed recommendations for LCA practitioners such as developing predictive scenario analyses or generic scale-up scenarios.

Although the integration of LCA into the development phase of products is rated as beneficial– for instance, when determining if the new solution is better for the environment than the currently available ones–approaches on how to address sustainability criteria in the early stage of product development are still limited (Tao et al. 2017). The collection of data is often seen as the most time-consuming part of an LCA (Baumann and Tillman 2004). Whereas the issue of data quality has been discussed frequently in the literature (e.g., Guinée 2002; Guo and Murphy 2012; United States Environmental Protection Agency 1995), very little has been published about how to carry out the data collection (Baumann and Tillman 2004; Kuczenski et al. 2018). The limitations diagnosed are understood as referring to practical recommendations on how to implement an LCA procedure in an R&D project. Considering that, the indications from literature so far present themselves as fragmented or not even represented in scientific scholarship because such information is part of the implicit knowledge of LCA practitioners.

While it seems that the concepts of bioeconomy and sustainable development have common goals (Bugge et al. 2016; European Commission 2012; McCormick and Kautto 2013; Van Lancker et al. 2016), the reputation of bio-based raw materials, such as wood, as being a sustainable material per se, is not enough for declaring the sustainability of bio-based products in general (Appelhanz et al. 2016; Clancy et al. 2013; Sherwood et al. 2017). Hence a systematic assessment of the potential environmental impacts of newly developed bio-based products along their life cycles is necessary. Aside of LCA, a vast number of approaches to integrating various sustainability considerations in the development of new products have been developed in order to support or enable informed decision-making in the context of sustainable development. The range of sustainability assessment approaches reaches from micro-to macro-scale, from procedural to analytical tools and from qualitative to quantitative approaches (Singh et al. 2012). The assessment of environmental impacts of newly developed bio-based products in R&D projects is challenged by various aspects, such as the need for impact categories beyond the emission- oriented ones, considering, for example, also land use change and carbon sinks as critical aspects (Cowell and Clift 1997; Pawelzik et al. 2013; Weiss et al. 2012). Although processes and products in R&D projects often aim to replace others, it is noteworthy that the direct comparison is often challenged by the comparability of the systems' functionalities itself, scale-up issues, technological uncertainties as well as performance uncertainties (Hetherington et al. 2014). Furthermore, conducting LCA for new products is often tricky due to low data availability (Hetherington et al. 2014). The integration of LCA into the R&D process can be beneficial for guiding the development of environmentally advantageous product systems (Arvidsson et al. 2018; Lettner et al. 2018; Hesser et al. 2017; Hetherington et al. 2014; Sandin et al. 2014). LCA, by its nature, is a comprehensive methodology but is often criticized for being too complex, time-consuming, expensive and sometimes even insufficient for R&D. Nevertheless, its capacity to guide sustainable product development justifies its implementation (Chang et al. 2014; Devanathan et al. 2010; Hetherington et al. 2014; Kunnari et al. 2009; Nielsen and Wenzel 2002; Piccinno et al. 2016; Sandin et al. 2014). Additionally, funding agencies are pushing towards the integration of environmental assessment in the early stages of product development, leading to a rethinking of process and product development. For instance, the European Commission encourages environmentally conscious R&D in publicly funded research programs, such as Horizon 2020 (Baldassarri et al. 2016). Paradoxically, R&D projects are dominated by technological progress, with a lack of considering other influences such as environmental aspects (Golembiewski et al. 2015; Roos et al. 2014). The need for a systemic integration of LCA in the R&D process is derived from the abovementioned. Therefore, the objective of this study is to provide practice-oriented recommendations for integrating LCA in R&D projects. A conceptual framework on how streamlined LCA can be carried out as an integral part of an ongoing R&D project to support a target-oriented R&D process will be showcased. The example case shows the valorization of a bio-based raw material into a bio-based product. The target audience of this study consists not only of LCA practitioners who aim to conduct LCA within R&D, but moreover of technical researchers and project managers interested in enhancing the environmental performance of their technical R&D projects by integrating LCA.

13.2 Methods

The practice-oriented recommendations for integrating LCA in R&D projects are derived from the experience and observations during an R&D project. The presented case is an example from a European Horizon 2020 project designed to contribute to the vision of bio-economy. The primary objective of the three-year inter-organizational and transdisciplinary project SmartLi (Smart Technologies for the Conversion of Industrial Lignin into Sustainable Materials) was to develop valorization routes for lignin, a by-product of the pulp and paper industry that is currently underutilized in terms of material use. Therefore, different pretreatment technologies were investigated to develop lignin-based products. The project consortium covered the value chain from pulp production to the final lignin-based products. The issue of environmental aspects was covered by a dedicated work package. The activities in the SmartLi project spanned between Technology Readiness Level (TRL, for details see EARTO 2014) 2 (technology concept formulated) and TRL 5 (technology validated in relevant environment).

Instead of a full LCA, which grasps the whole life cycle and all input and output flows, a streamlined LCA was carried out with a cradle-to-gate perspective, considering the impact categories Global Warming Potential (GWP) and the Cumulative Energy Demand (CED). The use of a screening approach to streamline the LCA at product development is one possibility to overcome challenges of LCA in R&D. Such challenges can be limitations in data availability due to the R&D phase which affects the life cycle inventory (e.g., definition of the system boundaries and functionalities of the novel product and processes), or also the characterization of potential environmental impacts of novel substances. Whereas Wenzel (1998) encourages to making use of the iterative nature of LCA for to the integration in R&D, Fleischer and Schmidt (1997) refer to screening LCA approaches as a requirement. While starting with a qualitative or semi-qualitative approach, Wenzel (1998) describes the screening LCA as mainly working with quantitative data. Wenzel's (1998) characterization of the basic levels of LCA, where a screening LCA is not yet as detailed as a full-scale LCA, can be considered as bottom-up characterization of LCA studies. In contrast, Klöpffer (2014) states that a streamlined LCA is reduced by elements that have been identified as not significantly affecting the results—such as certain input flows or a specific process or life cycle stage—for the sake of reducing its complexity. The streamlined approach is often considered as a starting point for further in-depth analysis, determining whether a more detailed LCA is needed or on which aspect to focus (Todd and Curran 1999). Fleischer and Schmidt (1997) emphasize this aspect by using the phrase "iterative screening LCA" for their approach to make use of LCA in relation to R&D. Each iteration step is finalized by feedback with the team of developers, which then becomes the basis for further refinement of the LCA model in terms of details considering system boundaries, data quality, assumptions, etc. (Fleischer and Schmidt 1997). They also conclude that the implementation of screening LCA approaches are essential for environmentally sound product development because the development stage is the most effective stage to reduce environmental impacts. The streamlined LCA of the presented R&D project was built upon the following principles:

- Integrated, referring to the LCA as part of the innovation management by establishing feedback loops from technology development to environmental assessment and vice versa;
- Iterative, referring to the refining of the LCA study, along with R&D advancements when new information is available;
- Modular, referring to grouping the assessment of unit processes and connecting those to different process pathways depending on the R&D advancements.

Figure 13.1 shows the schematic procedure of the LCA as conducted in the SmartLi project (see detailed descriptions indicated by letters from A to K below).

As part of the inter-organizational project comprising representatives of academia, industry and an association, the first step was to provide background information about LCA, which includes the clarification of the role and aim of the LCA within

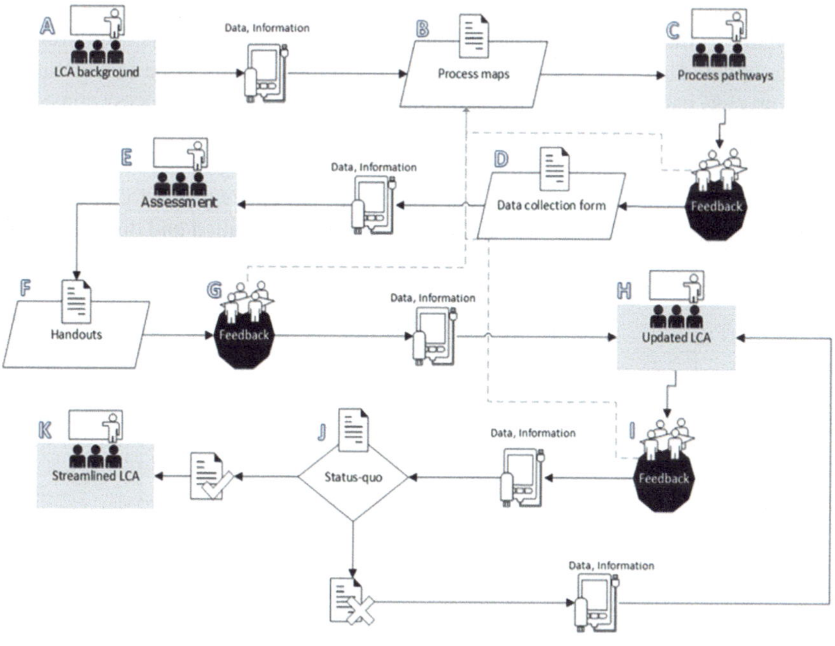

Fig. 13.1 Schematic procedure of setting up and conducting an LCA within an R&D project at low TRL

the project (see A in Fig. 13.1). According to Sandin et al. (2014), there are eight roles an LCA can have in technical R&D processes:

- Guide technical R&D;
- Develop life-cycle thinking;
- Support scale-up;
- Direct future R&D activities;
- Market technology;
- Demonstrate inclusion of environmental concerns;
- Contribute to LCA knowledge;
- And/or fulfil requirements (of commissioner or funding agency).

As stated by Sandin et al. (2014), the role of the LCA described by an LCA practitioner does not have to be the same role as described by other project partners. Typical for LCA in early R&D stages is the aim to provide information for various stakeholders, i.e., project partners (Hetherington et al. 2014). In terms of the SmartLi project, one role was explicitly stated in the project proposal: "A Life Cycle Assessment should be carried out in order to evaluate the environmental and socio-economic performance of the developed technologies". For the LCA practitioners, one additional aim was to contribute to LCA knowledge in the field of LCAs in R&D projects.

After gathering information about the unit processes from the project partners, the process map was developed (see B in Fig. 13.1). This initial process map contained all of the potential process pathways the project partners could come up with, from Kraft lignin to the target lignin-based polyurethane and epoxy resins. The process map was updated iteratively with further project progress, leading to the consideration of new process pathways or exclusion of others based on chemical process considerations (see Fig. 13.2). Additionally, the process map was designed to already reflect potential uncertainties, meaning that process pathways with high risk of development failure, such as achieving desired product properties were shown as dotted lines. The potential process pathways from the process map were presented and discussed at a

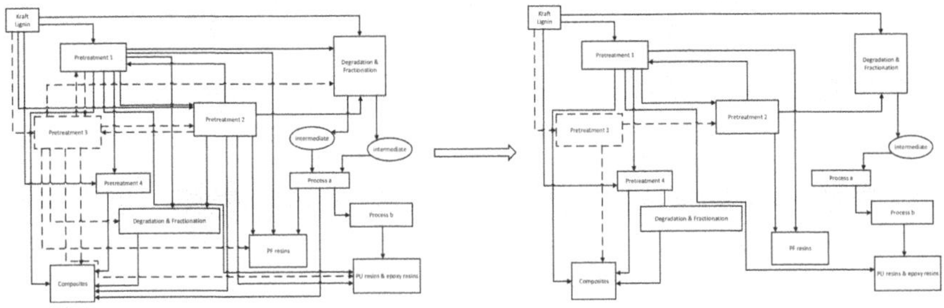

Fig. 13.2 Process map containing potential process pathways from Kraft lignin to polyurethane (PU) and epoxy resins via different pretreatment technologies. Dotted lines indicate pathways with high development risk. The variety of possible process pathways (left) was narrowed down during the development process (right)

project meeting (see C in Fig. 13.1). Because of the high number of potential process pathways, the consortium agreed to first have a closer look at the most likely process pathway for each of the targeted lignin-based products.

When the LCA practitioner does not personally collect the data, it is important to provide an intuitive and easy-to-handle document for the person conducting the data collection. Therefore, the next step included the preparation of a simple data collection form for all the unit processes within the process pathways agreed on beforehand (see D in Fig. 13.1). The data collection form was provided as electronic calculation tables, containing a short explanation about LCA as well as questions on the process name, a short process description, date of data collection and a contact person. This ‘general’ section of the form helped the LCA practitioners to familiarize themselves with the technical vocabulary.

The data collection form was further structured into the following sections:

- Inputs: raw materials and chemicals; water; auxiliaries; and energy inputs;
- Outputs: products and by-products;
- Waste and emissions to air, water and land;
- Additionally, a section for any kind of remarks or comments was included.

In line with the iterative process, each data collection form was updated regularly (e.g., after process change or scale-up) in line with the R&D advancements. The results of the assessment were then presented to all project partners (see E in Fig. 13.1) to build a common level of information and exemplify how the collected data was processed. Additionally, handouts with detailed results and a preliminary hot-spot analysis for the target products were prepared and distributed (see F in Fig. 13.1). Following the principle of an integrated approach, based on the results of the assessment a two-way feedback loop (see G in Fig. 13.1) was established:

- Feedback from LCA practitioners to project partners can address information on first as well as revised results, potential environmental hot-spots, explanation of next steps for the assessment, further data requests, challenges, etc;
- Feedback from project partners to LCA practitioners can address changes in process and/or product development, changes of process pathways, data availability, changes in interest, etc.

After each feedback round, the LCA practitioners are required to evaluate their assessments accordingly. This might also include the adaptation of individual unit processes, leading to the redesign of the process map (see Fig. 13.2). Followed by presenting and discussing the updated assessment (see H and I in Fig. 13.1). This feedback from the project partners on the process pathways was again integrated and the process maps were again updated at this point (see H and I in Fig. 13.1) (see Fig. 13.2). Throughout the R&D project the LCA was iteratively updated and then presented to the project partners and followed again by a feedback loop from B to I in Fig. 13.1. The status quo (see J in Fig. 13.1) represents the step where consensus was sought before communicating the LCA results outside of the project consortium. If all involved stakeholders agreed on the current status of the LCA (indicated by the check in Fig. 13.1), the streamlined LCA was considered ready for presentation and publication (see K in Fig. 13.1). If not, the feedback loop would be continued in order to update the LCA until consensus was reached—which might involve further R&D. It should be mentioned that the presented procedure may differ in the individual steps depending on the project.

13.3 Results

In the following, recommendations for integrating LCA in technical R&D projects derived from the presented example case are presented. This section is structured into the following three recommendations.

13.3.1 Create Common Knowledge

When condicting LCA at early research stages it is recommended to frame its primary objective to support R&D. This can be done by generating and integrating information for the further development process rather than mandate the technological partner how to develop an environmentally friendly innovation. If LCA practitioners tried to force technical R&D in such a way, this might lead to tensions within the project consortium. LCA can generate, process and provide information on the environmental performance of the newly developed products under the given assumptions. With this information, environmental hotspots can be anticipated and encountered proactively by integrating them into the further development agenda. Hence, LCA does not narrow down the technical research but reveals improvement and development potentials. This requires that the technical research partners engage in scientific collaboration with the LCA practitioners rather than only consulting these experts.

In order to take full advantage of the potential to support ongoing R&D with LCA, it was observed that it is necessary to provide LCA background information to the project partners. Additionally, the role, the goal and the benefits of the LCA should be clarified before starting any data collection (i.e., for the life cycle inventory). Background information may consist of exemplary results from other projects and, to a certain extent, of elaborating on the basic LCA method and its logic. The intensity of this step will essentially depend on the prior knowledge of the project partners. The role of LCA in the R&D project should be clarified in order to lay out the interests and identify opportunities for the project partners in order to enhance commitment. Whereas for LCA practitioners the role of contributing to LCA knowledge might be obvious, the technical project partners might wish for other roles, such as marketing purposes. Which role or roles the LCA can have strongly depends on the project and progress itself.

Within the described case, multiple roles evolved out of the initially defined setting in the project proposal. Although a change in roles was experienced, it is recommended to clarify the role of LCA together with the project partners as a first common step towards determining all patries' expectations. Along with clarifying what the LCA practitioner is expected to contribute to the R&D project, it needs to be clarified how the data collection will be done, how the data is processed, used and made available in the project consortium. It is also suggested to clarify how (i.e. presentation, project reports, etc.) and how often the LCA practitioners communicate (preliminary) LCA results.

Following the proposed principle of integrating LCA into an R&D project, experience revealed that developing a common language and feedback mechanism is the key towards integrating LCA in technical R&D projects to support a target-oriented process and product development. As in all interdisciplinary project collaborations, communication is a major success factor. In this case, it was experienced that the data collection, served as tool to build a common language. Whereas for well-established processes and products data can be quite easily found in LCA databases (e.g., Ecoinvent, Gabi, etc.), primary data need to be collected when it comes to new processes or products. In this case, the project partners participated actively in the data collection. This not only supported the finding of common language and terminology between the project partners, it also increased willingness to share data and accept the data analysis. Referring to Baumann and Tillman (2004), it is recommended to ask the project partners to additionally provide information on properties of in- & outputs (e.g., solid stage, concentration, purity, energy source etc.), and already at early stages (lab scale) and on potential up-scaling effects (e.g., changing solvents, energy needs, additional process steps, recovery and recycling processes, wastewater treatment, etc.). The descriptive and qualitative data provided by the technical project partners help the LCA practitioners to better understand the technical terminology and integrate the LCA into the R&D.

13.3.2 Consider LCA as Part of the R&D

Following the principle of integration, it is recommended to establish periodical presentations on the current LCA status (i.e., at the regular project meetings) and link those presentations to feedback possibilities and a data collection schedule to update the LCA. Communicating the results by providing handouts for each target product (i.e., in our case that meant four products) to the project partners has proven to be an effective way to build a common understanding of the R&D advancements in terms of environmental issues. The aim for providing handouts to the project partners was not only to present preliminary results clearly to non-LCA practitioners but also to trigger the next round of data collection. The two-way feedback loops from LCA practitioner to project partner and vice versa is recommended for a proactive communication culture.

How to conduct these feedback loops depends on the nature of the project. The observations within the case study revealed that it was helpful to receive both oral and a written feedback. The first was collected in order to encourage the discussion of environmental issues in the project. The second feedback included a written communication providing the opportunities for the project partners to ask questions regarding the LCA. The partners were asked: "What would you like to know regarding your process/product in terms of LCA?" It is assumed that by explicitly asking the project partners that not only communication and interest would be enhanced but also the general support for the integration of the LCA. Exemplary answers were as follows: "What will change if we change input X to input Y?" or "In comparison with product X, how is the environmental performance of our product Y?" Also, before the results of the LCA were communicated to a broader audience (e.g., as part of project reports), a feedback loop on the status-quo was carried out. Any discrepancies should then be clarified and, if necessary, new data and/or information should be collected. The results should be updated again accordingly in order to develop a consensus about the results published. This is a simple approach to improving communication as well as enhancing the project partners' commitment to the LCA results. This ensures that the information actually required for the target-oriented process and product development is obtained. It is recommended to encourage LCA practitioners in R&D projects to develop a participatory momentum.

In the presented case, the LCA practitioners were able to observe that, as the project continued, the data availability increased while uncertainties simultaneously decreased. For example, whereas it was difficult to assess the energy demand at the beginning of the project (lab scale, low TRL), with increased TRL, the project partners were able to provide data on measured energy consumption. Additionally, as the possible up-scaling effects became more precise, it was possible to integrate those assumptions into the LCA, for example, by simulating a recovery process of used chemicals (see also the case study by Lettner et al. 2018) Thus, it was possible to provide various scenarios and indicate how the environmental performance of the processes and products can be improved. This kind of information strongly supported the LCA as an important part of the R&D rather than a side-effect.

13.3.3 Adapt and Revise

It is recommended to make full use of the iterative feature when integrating LCA in R&D projects, because changes in potential environmental impacts related to R&D advancements are reflected systematically and when needed. Thereby the technical project partners are able to perceive/identify the impacts of their decisions considering processes and inputs (which again creates commitment). It was observed that it is important to explain and stress, especially to the non-LCA practitioners in the project, the iterative nature of LCA. All project partners should be aware that the model of the considered processes (i.e., system boundaries, functional unit, etc.) as well as the results of the LCA will change in line with the R&D advancements.

It was observed, that in light of a new product under development also the definition of the associated value chain is uncertain. Followed by a continuous restructuring of the process map and the availability of new information, the modular approach has proven to be very practical. By implementing a modular approach instead of modelling all alternatives, in a first step the individual unit processes were assessed. In accordance with the project partners, those individual unit processes were then linked to new process pathways. However, it is noteworthy, that special attention must be paid to ensuring that the respective inputs and outputs fit together to enable new process pathways. The modular approach thus provides the opportunity to accomplish the following: (a) assess each module by itself (e.g., hot-spot analysis within one system), (b) easily add new information (e.g., up-scaling of developed technologies), and (c) support the development of the LCA parallel to the technological advancements.

In order to support the ongoing R&D, a periodical update on resource input and process data for the life cycle inventory and following the LCA is required. In line with the iterative approach, each data collection file should be updated regularly. A regular update further supports an integrated LCA routine, meaning that the project partners are aware of the planned data collection schedule. Data collection should be performed according to a time schedule, but, especially in R&D, the data collection should be additionally carried out in line with R&D advancements (e.g., after process change or scale-up). Hence, deviations from the data collection plan may occur, which requires a high degree of flexibility on the LCA practitioners' part. Whereas the data collection is usually triggered by an explicit request from the LCA practitioners, in well-integrated LCA the data collection and the information flow should also be triggered by the project partners.

Early stage technical research is a phase of trial and error. That is why data inquiry might be perceived as inadequate, pushy or controlling. Implementing iterations in the LCA therefore also leads to pressure both the LCA practitioners and the technical project partners exert, considering the R&D and the LCA results because the results are preliminary up to the point when consensus on a certain status quo of the LCA is reached. The iterative approach also may entail the general issue of not having adequate data for conducting LCA of product systems under development. The initial qualitative process mapping can be developed towards a quantified LCA model and assumptions, proxies, analogy data, data from different processing or maturity scales amongst others can be refined iteratively. Furthermore, another typical issue that was observed is, that during the early research stage the exact functional parameters (for instance stiffness, elastic modulus, tensile strength,…) of the considered product cannot be defined yet, often leading to a comparison based on masses or volumes between products instead of a comparison based on functions. In such a case it is recommended to operate with bandwidths (e.g. derived from literature and scenarios) of potential environmental impacts and opportunity spaces—instead of applying the best estimate principle—to enhance the understanding of the product system and to define at least one process pathway for each target product and if possible an additional one for comparison. This in turn creates motivation to reach best case scenarios and thus supports the aim to reduce environmental impacts.

13.4 Discussion and Conclusion

The use of LCA and other sustainability assessment methods during R&D are discussed controversially in literature. Whereas most of the studies suggest a streamlined approach (Fleischer and Schmidt 1997; Hur et al. 2005; Pascual-González et al. 2015; Todd and Curran 1999), others (Millet et al. 2007) indicate the limitations of LCA in product development. This study intended to provide practice-oriented recommendations for integrating LCA in R&D at early stages for LCA practitioners and non-LCA practitioners. Whether LCA is a suitable method for a specific project or not should be decided on a case-by-case basis and cannot be answered in general. Nevertheless, LCA is considered to be the best-developed methodology in the area of sustainability assessment (Finnveden et al. 2009). Although it seems that there is a push towards integrating environmental aspects in R&D projects by funding authorities, technologically driven project management often lacks in systematically considering the environmental—and also social—dimension. Consequently, questions remain unanswered on how to apply and integrate LCA in R&D projects (Baldassarri et al. 2016; Brones et al. 2014; Dufrene et al. 2013). While upon first consideration LCA may seem too complex and time-consuming to be applied in R&D projects, it is concluded from the observations of this case that by taking the advantage of the iterative character of the method and implementing a conscious communication, LCA can build capacities to support the technical-driven R&D towards an environmental conscious process and product development. LCA can be used as innovation management tool to support technical R&D projects by proactively developing knowledge on potential environmental impacts in line with the R&D advancements and reducing development risks at the same time.

Sandin et al. (2014) identified eight roles an LCA can have in technical R&D processes. As it is with the general suitability of LCA also the roles of LCA are case-dependent. We agree with Sandin et al. (2014), if the roles are not pre-defined (e.g., in the project proposal) the early definition of the role in the project supports its fulfilment. Additionally, it needs to be pointed out that the role(s) can change over the project progress. In our case study, we were able to observe two roles at the beginning of the project: Firstly, the role of fulfilling the requirements by the funding agency and secondly developing a life-cycle thinking within the project. In line with the project progress all of the other roles described by Sandin et al. (2014) were more or less encountered. LCA practitioners should be aware of this potential shift in expectations from the project partners regarding the LCA. This does require the mentioned regular communication and information sharing between the LCA practitioners and the project partners. It is not only the task of the LCA practitioner to decide how this communication and information sharing is carried out (e.g., regular meetings, handouts, etc.). It is also his/her responsibility to decide how to communicate the LCA results so that all stakeholders clearly understand data provided to them (Hetherington et al. 2014). LCA practitioners should be aware that, especially in early research stages, not only the LCA has to deal with uncertainties but those early stages are said to be based on "trail and error" of the technical research is a phase of 'trial and error'. Sandin et al. (2014), for instance, observed that tensions can occur because LCA practitioners request data that are not yet available at the early stage of the project. Another aspect in line with communication is the issue of data and knowledge management as well as confidentiality issues potentially limiting transparency (Kuczenski et al. 2018). As R&D processes are highly complex processes sharing knowledge among the R&D partners can avoid redundancy and allow the diffusion of best practices (Huang 2009). We thus spot another possible role that LCA can have in technical R&D projects especially in inter-organizational and transdisciplinary projects, LCA can be seen as a connection point between the project partners along the value chain besides the pure technical exchange and thus provide a communication platform. Consequently, it is concluded that, especially in early research stages, LCA can contribute to an increase in communication among the project partners by providing a common platform and thus bridge potential barriers in knowledge sharing among technical project partners by presenting technical results translated into environmental impacts. Several comments or findings presented in the introductory literature were also observed in this case. The LCA practitioners have to be clear and consistent in using terms and concepts. Heijungs (2014) indicated, for instance, the importance of using technical terms correctly, which may differ among different scientific disciplines. Developing the data collection together in the project consortium across disciplines was helpful to overcome this pitfall. Additionally, it is recommended that LCA practitioners should be aware of the prior LCA knowledge, experience and expectations their project partners might have. For instance, Testa et al. (2016) conducted a study on the perception of LCA implementations among adopters and non-adopters, showing that the perceived benefits and barriers of LCA differ depending on the prior experiences. Although in scientific literature the most common benefit of LCA is seen for the environmental improvement of products or services, the study among companies showed that other aspects, such as being a tool to drive strategic decisions, are seen as benefit when implementing LCA (Testa et al. 2016). In conclusion how and how strongly the LCA can be integrated into a project of course depends on many issues, from the general goal of the project, the project consortium, the duration of the project, but also of course on the experience and motives of the LCA practitioners. Universal validity of the presented recommendations is not claimed. Further testing and adaptations of the conceptual framework are required to increase the robustness of the approach described and to understand further needs of technical R&D projects in terms of environmental considerations.

Acknowledgements This work was partially funded by the SmartLi project under Horizon 2020/BBI-JU, the Austrian research promotion agency (FFG) under the scope of the COMET program (wood comet 865905).

References

Albrecht S, Brandstetter P, Beck T, Fullana-i-Palmer P, Grönman K, Baitz M, Deimling S, Sandilands J, Fischer M (2013) An extended life cycle analysis of packaging systems for fruit and vegetable transport in Europe. Int J Life Cycle Assess 18:1549–1567

Appelhanz S, Osburg V-S, Toporowski W, Schumann M (2016) Traceability system for capturing, processing and providing consumer-relevant information about wood products: system solution and its economic feasibility. J Clean Prod 110:132–148

Arvidsson R, Tillman AM, Sandén BS, Janssen M, Nordelöf A, Kushnir D, Molander S (2018) Environmental assessment of emerging technologies: recommendations for prospective LCA. J Ind Ecol 22(6):1286–1294

Baldassarri C, Mathieux F, Ardente F, Wehmann C, Deese K (2016) Integration of environmental aspects into R&D inter-organizational projects management: application of a life cycle-based method to the development of innovative windows. J Clean Prod 112(4):3388–3401

Baumann H, Tillman AM (2004) The Hitch Hiker's Guide to LCA: an orientation in life cycle assessment methodology and application. Lund, Studentlitteratur

Brones F, de Carvalho MM, de Senzi Zancul E (2014) Ecodesign in project management: a missing link for the integration of sustainability in product development? J Clean Prod 8:106–118

Bugge MM, Hansen T, Klitkou A (2016) What is the bioeconomy? A review of the literature. Sustainability 8:691

Chang D, Lee CKM, Chen C-H (2014) Review of life cycle assessment towards sustainable product development. J Clean Prod 83:48–60

Cherubini F, Bargigli S, Ulgiati S (2009) Life cycle assessment (LCA) of waste management strategies: landfilling, sorting plant and incineration. Energy 34:2116–2123

Clancy G, Fröling M, Svanström M (2013) Changing from petroleum to wood-based materials: critical review of how product sustainability characteristics can be assessed and compared. J Clean Prod 39:372–385

Cowell SJ, Clift R (1997) Impact assessment for LCAs involving agricultural production. Int J Life Cycle Assess 2:99–103

Danilecki K, Mrozik M, Smurawski P (2017) Changes in the environmental profile of a popular passenger car over the last 30 years—results of a simplified LCA study. J Clean Prod 141:208–218

Devanathan S, Ramanujan D, Bernstein WZ, Zhao F, Ramani K (2010) Integration of sustainability into early design through the function impact matrix. J Mech Design, Trans ASME 132:0810041–0810048

Ding N, Gao F, Wang Z, Yang J (2016) Life cycle energy and greenhouse gas emissions of automobiles using aluminum in China. J Ind Ecol 20:818–827

Dufrene M, Zwolinski P, Brissaud D (2013) An engineering platform to support a practical integrated eco-design methodology. CIRP Ann 62:131–134

EARTO (2014) The TRL scale as a research & innovation policy tool. EARTO Recommendations. 30.04.2014, acessed 19.01.2020 under https://www.earto.eu/our-publications/

European Commission (2012) Innovating for sustainable growth: a bioeconomy for Europe. European Commission, Brussel, Belgium

European Commission (2019) European platform on life cycle assessment (LCA). 07.08.2019, acessed 11.01.2020 under https://ec.europa.eu/environment/ipp/lca.htm

Finnveden G, Hauschild MZ, Ekvall T, Guinée J, Heijungs R, Hellweg S, Koehler A, Pennington D, Suh S (2009) Recent developments in life cycle assessment. J Env Manag 91:1–21

Fleischer G, Schmidt W-P (1997) Iterative screening LCA in an eco-design tool. Int J Life Cycle Assess 2:20–24

Golembiewski B, Sick N, Bröring S (2015) The emerging research landscape on bioeconomy: what has been done so far and what is essential from a technology and innovation management perspective? Innov Food Science Emerg Technol 29:308–317

González-García S, Berg S, Moreira MT, Feijoo G (2009) Evaluation of forest operations in Spanish eucalypt plantations under a life cycle assessment perspective. Scand J Forest Res 24:160–172

Guinée J (2002) Handbook on life cycle assessment: operational guide to the ISO Standards. Kluwer Academic Publishers, Dordrecht

Guo M, Murphy RJ (2012) LCA data quality: sensitivity and uncertainty analysis. Sci Tot Env 435(436):230–243

Heijungs R (2014) Ten easy lessons for good communication of LCA. Int J Life Cycle Assess 19:473–476

Hesser F, Wohner B, Meints T, Stern T, Windsperger A (2017) Integration of LCA in R&D by applying the concept of payback period: case study of a modified multilayer wood parquet. Int J Life Cycle Assess 22(3):307–316

Hetherington AC, Borrion AL, Griffiths OG, McManus MC (2014) Use of LCA as a development tool within early research: challenges and issues across different sectors. Int J Life Cycle Assess 19:130–143

Huang C-C (2009) Knowledge sharing and group cohesiveness on performance: an empirical study of technology R&D teams in Taiwan. Technovation 29:786–797

Hur T, Lee J, Ryu J, Kwon E (2005) Simplified LCA and matrix methods in identifying the environmental aspects of a product system. J Env Manage 75:229–237

Karjalainen T (2001) Energy, carbon and other material flows in the life cycle assessment of forestry and forest products: achievements of the working group 1 of the COST action E9. European Forest Inst, Joensuu

Klöpffer W (2014) Background and future prospects in life cycle, vol 19. Springer, Berlin Heidelberg. https://doi.org/10.1007/s11367-014-0798-y

Kuczenski B, Marvuglia A, Astudillo MF, Ingwersen WW, Satterfield MB, Evers DP, Koffler C, Navarrete T, Amor B, Laurin L (2018) LCA capability roadmap—product system model description and revision. Int J Life Cycle Assess 23:1685–1692

Kunnari E, Valkama J, Keskinen M, Mansikkamäki P (2009) Environmental evaluation of new technology: printed electronics case study. J Clean Prod 17:791–799

Lacasa E, Santolaya JL, Biedermann A (2016) Obtaining sustainable production from the product design analysis. J Clean Prod 139:706–716

Laurent A, Bakas I, Clavreul J, Bernstad A, Niero M, Gentil E, Hauschild MZ, Christensen TH (2014) Review of LCA studies of solid waste management systems—part I: lessons learned and perspectives. Waste Manag 34:573–588

Lettner M, Solt P, Rößiger B, Pufky-Heinrich D, Jääskeläinen A-S, Schwarzbauer P, Hesser F (2018) From wood to resin—identifying sustainability levers through hotspotting lignin valorisation pathways. Sustainability 10:2745

Luz LMD, Francisco ACD, Piekarski CM, Salvador R (2018) Integrating life cycle assessment in the product development process: a methodological approach. J Clean Prod 193:28–42

McAloone TC, Pigosso DCA (2017) From ecodesign to sustainable product/service-systems: a journey through research contributions over recent decades. In: Stark R, Seliger G, Bonvoisin J (eds) Sustainable Manufacturing. Sustainable production, life cycle engineering and management. Springer, Cham

McCormick K, Kautto N (2013) The bioeconomy in Europe: an overview. Sustainability 5:2589–2608

McManus MC, Taylor CM (2015) The changing nature of life cycle assessment. Bioma Bioenerg 82:13–26

Meisterling K, Samaras C, Schweizer V (2009) Decisions to reduce greenhouse gases from agriculture and product transport: LCA case study of organic and conventional wheat. J Clean Prod 17:222–230

Millet D, Bistagnino L, Lanzavecchia C, Camous R, Poldma T (2007) Does the potential of the use of LCA match the design team needs? J Clean Prod 15:335–346

Nielsen PH, Wenzel H (2002) Integration of environmental aspects in product development: a stepwise procedure based on quantitative life cycle assessment. J Clean Prod 10:247–257

Pascual-González J, Pozo C, Guillén-Gosálbez G, Jiménez-Esteller L (2015) Combined use of MILP and multi-linear regression to simplify LCA studies. Comput Chem Engin 82:34–43

Pawelzik P, Carus M, Hotchkiss J, Narayan R, Selke S, Wellisch M, Weiss M, Wicke B, Patel MK (2013) Critical aspects in the life cycle assessment (LCA) of bio-based materials—reviewing methodologies and deriving recommendations. Res Cons Recyc 73:211–228

Piccinno F, Hischier R, Saba A, Mitrano D, Seeger S, Som C (2016) Multi-perspective application selection: a method to identify sustainable applications for new materials using the example of cellulose nanofiber reinforced composites. J Clean Prod 112:1199–1210

Piringer G, Bauer A, Gronauer A, Saylor M, Stampfel A, Kral I (2016) Environmental hot spot analysis in agricultural life-cycle assessments–three case studies. J Centr Europ Agric 17:477–492

Roos A, Lindström M, Heuts L, Hylander N, Lind E, Nielsen C (2014) Innovation diffusion of new wood-based materials—reducing the "time to market". Scand J Forest Res 29:394–401

Sandin G, Clancy G, Heimersson S, Peters GM, Svanström M, ten Hoeve M (2014) Making the most of LCA in technical inter-organisational R&D projects. J Clean Prod 70:97–104

Scherhaufer S, Moates G, Hartikainen H, Waldron K, Obersteiner G (2018) Environmental impacts of food waste in Europe. Waste Manag 77:98–113

Schöggl J-P, Baumgartner RJ, Hofer D (2017) Improving sustainability performance in early phases of product design: a checklist for sustainable product development tested in the automotive industry. J Clean Prod 140:1602–1617

Sempels C, Hoffmann J (2013) Sustainable innovation strategy—creating value in a world of finite resources. Palgrave Macmillan, UK. ISBN 978-1-137-35260-6

Sherwood J, Clark JH, Farmer TJ, Herrero-Davila L, Moity L (2017) Recirculation: a new concept to drive innovation in sustainable product design for bio-based products. Molec 22:48

Singh RK, Murty HR, Gupta SK, Dikshit AK (2012) An overview of sustainability assessment methodologies Ecol Indic 15:281–299

Siracusa V, Ingrao C, Lo Giudice A, Mbohwa C, Dalla Rosa M (2014) Environmental assessment of a multilayer polymer bag for food packaging and preservation: an LCA approach. Food Res Int 62:151–161

Tao J, Chen Z, Yu S, Liu Z (2017) Integration of life cycle assessment with computer-aided product development by a feature-based approach. J Clean Prod 143:1144–1164

Testa F, Nucci B, Tessitore S, Iraldo F, Daddi T (2016) Perceptions on LCA implementation: evidence from a survey on adopters and nonadopters in Italy. Int J Life Cycle Assess 21:1501–1513

Todd JA, Curran MA (1999) Streamlined life-cycle assessment: a final report from the SETAC North America Streamlined LCA Workgroup

Trappey AJC, Ou JJR, Lin GYP, Chen M-Y (2011) An eco- and inno-product design system applying integrated and intelligent qfde and triz methodology. J Syst Science Syst Engin 20:443–459

Unger N, Beigl P, Höggerl G, Salhofer S (2017) The greenhouse gas benefit of recycling waste electrical and electronic equipment above the legal minimum requirement: an Austrian LCA case study. J Clean Prod 164:1635–1644

United States Environmental Protection Agency (1995) Guidelines for assessing the quality of life cycle inventory analysis. In: U.S.E.P. Agency (ed)

Van Lancker J, Wauters E, Van Huylenbroeck G (2016) Managing innovation in the bioeconomy: an open innovation perspective. Biom Bioenerg 90:60–69

Weiss M, Haufe J, Carus M, Brandão M, Bringezu S, Hermann B, Patel MK (2012) A review of the environmental impacts of biobased materials. J Ind Ecol 16:169–181

Wenzel H (1998) Application dependency of LCA methodology: key variables and their mode of influencing the method. Int J Life Cycle Assess 3:281–288

Wikström F, Williams H, Venkatesh G (2016) The influence of packaging attributes on recycling and food waste behaviour—an environmental comparison of two packaging alternatives. J Clean Prod 137:895–902

Zafeirakopoulos IB, Genevois ME (2015) An analytic network process approach for the environmental aspect selection problem—a case study for a hand blender. Env Impact Asses Rev 54:101–109

Zah R, Hischier R, Leão AL, Braun I (2007) Curauá fibers in the automobile industry—a sustainability assessment. J Clean Prod 15:1032–1104

Chapter 14
Carbon Offsets: An LCA Perspective

Rosalie Arendt, Vanessa Bach, and Matthias Finkbeiner

Abstract Carbon offsets as an additional measure to mitigate climate change are on the agenda in recent years. This study analyzes the three carbon offsetting programs (the Clean Development Mechanism, the Verified Carbon Standard and the Gold Standard) with the largest market shares by systematically comparing their standard documents with environmental Life Cycle Assessment (LCA) standards (ISO 14067 and ISO 14040/44). The programs' most important methodologies are assigned to the sectors forestry, renewable energy, energy efficiency, industrial gas, and waste. We analyzed each sector for its compatibility with LCA using a criteria evaluation scheme to answer the main question, whether the methodologies provide guidance on life cycle emission accounting and what uncertainties they face. The offsetting standards differ from LCA standards due to different analyzed systems, system boundaries and purposes of their methods. Furthermore, offsetting methods always apply scenario analysis. Environmental impacts apart from greenhouse gases are not quantified, rather environmental impact assessments of heterogeneous quality are applied. We find that the approaches in the analyzed carbon offsets are incompatible with the LCA approach, mainly because they always involve scenario analysis, do not include all life-cycle phases and do not account for additional (negative) environmental and social impacts that project activities related to carbon offsets may cause.

Keywords Carbon offsets · Sustainable development · LCA · Climate change · Emission trading · Environmental accounting · Climate protection projects

14.1 Introduction

The climate crisis threatens the intactness of the world's ecosystems and the quality of human life (Masson-Delmotte 2018). Different policy approaches are being discussed to tackle the climate crisis, such as a carbon tax, emission trading and

R. Arendt (✉) · V. Bach · M. Finkbeiner
Technische Universität Berlin, Chair of Sustainable Engineering, Straße des 17. Juni 135, 10623 Berlin, Germany
e-mail: arendt@tu-berlin.de

S. Albrecht et al. (eds.), *Progress in Life Cycle Assessment 2019*,
Sustainable Production, Life Cycle Engineering and Management,
https://doi.org/10.1007/978-3-030-50519-6_14

environmental offsets (Aldy and Stavins 2011). Environmental offsets are a policy approach which intends to internalize the external costs of environmental damages to an economy at a minimum price (Pearce 2002). Offsets are permits that are generated by projects that aim at reducing pollutants (like implementation of a filter in industry) or enhancing the state of the environment (like the renaturation of a wetland). They are used on a voluntary (e.g. to offset emissions of flights) and on a compliance (e.g. European emission trading system) basis, depending on whether the purchase of a permit is required by law or not.

The most prominent carbon offsetting program was implemented through the Kyoto Protocol's Clean Development Mechanism (CDM), which created the world largest compliance market (Yamin 2014). The CDM publishes so-called methodologies for climate protection projects as well as tools and rules that contain additional information for their implementation. Thus, many methodologies for climate protection projects that can generate carbon[1] offsets have been established, of which 215 are still active today (UNFCCC, CDM 2019a). When these methodologies are applied, validated and certified by a certifier, the project developer can issue and sell Certified Emission Reductions (CERs) (UNFCCC, CDM 2019b). The CDM methodologies and their issuance and registration processes are the basis of voluntary carbon offsetting programs like the Verified Carbon Standard (VCS) and the Gold Standard, that hold 68 and 17% of the voluntary carbon market, respectively (Hamrick and Gallant 2017).

Existing analyses have shown that most projects registered under the CDM would have been realized even without money from the CDM (Schneider 2007; Cames et al. 2016). A similar weakness has been identified for the voluntary offsetting standards that are oriented towards the CDM standard and thus contain the same shortcomings (The Gold Standard 2019; Seyller et al. 2016). Deficiencies in the offsetting programs have been identified for the concept of additionality (that the realization of a project is strictly attributable to the money the project developers receive from the CDM), validation, verification and certification (Schneider and Möhr 2009; Kollmuss et al. 2008). While these studies focused on the additionality and certification quality of the projects, they did not focus on the methodology documents, their life cycle emission accounting, and possible trade-offs.

The CDM has already been criticized considering the life-cycle perspective: Böhm suggests that LCA might be a more suitable tool to calculate carbon emissions of climate protection projects (Böhm 2009). The ISO standard for Carbon Footprint (ISO 14067) states that offsetting is outside of the standard's scope. However, since some LCA scholars have already addressed the potential combination of offsets and LCA (Martínez-Blanco et al. 2015; Hellweg and Milà i Canals 2014), a systematic comparison of the two methods could clarify as to how they are compatible and where adjustments are necessary.

[1] Within this work, carbon is used as an equivalent to greenhouse gases and increased climate impact. A carbon offset is thus a commodity that claims to reduce a quantifiable amount of climate impact (increase of radiative forcing).

Thus, the aim of this study is to analyze carbon offsets from an LCA perspective, focusing on the respective documents (the ISO Standards for LCA and the complex methodologies and tools standards of the offsetting programs). To assess additional environmental aspects as well as social effects of the established projects additional literature, especially case studies from projects were used.

In the methods section, the used methods to analyze the carbon offsets from an LCA perspective are explained. The results section begins with a generic comparison of the ISO Standard for LCA and the offsetting programs that is summarized Table 14.2. Next, the relevant methodologies were determined by issuance share and assigned to sectors as displayed in Fig. 14.2. Each sector is then analyzed in more detail in its own subsection. The results are summarized in Table 14.3. Finally, the method and results are discussed and conclusions were drawn.

14.2 Method

An overview of the developed approach is displayed in Fig. 14.1. It shows the steps taken to analyze carbon offsets from an LCA perspective.

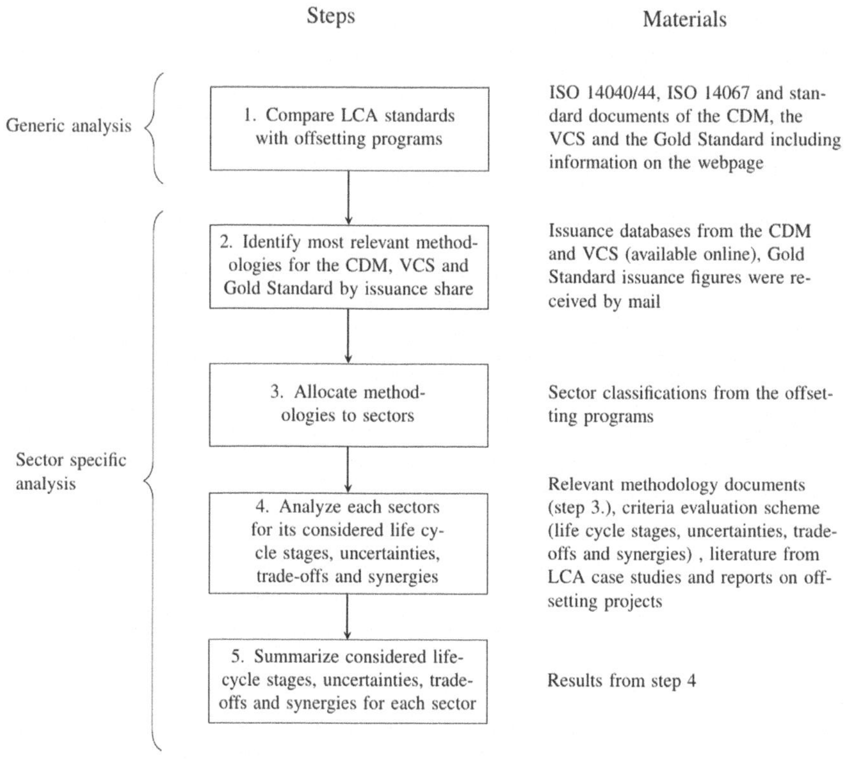

Fig. 14.1 Summary of method procedure

Table 14.1 Applied criteria to compare carbon offsets and LCA

Criteria	Guiding question
1. Analyzed system	Is the analyzed system a product or project?
2. Scenario	Is scenario analysis obligatory?
3. Life cycle phase	Does the method usually include all life cycle phases?
4. Output	What is the studies' output? Information? A commodity?
5. Review	Is a review obligatory and how is it performed?
6. Data collection	How flexible is the data collection and how is it performed?
7. System boundaries	What are the system boundaries and how are they defined?
8. Treatment of shift of burden and impact categories	How many impact categories are considered? What is done to avoid shift of burden?
9. Aggregation of biogenic and fossil carbon	Should biogenic and fossil carbon be aggregated or shown separately?

The core piece of the analysis is step one: *the comparison of the LCA standards with the carbon offsetting programs* and to underline their differences in a table (see Table 14.2). We derived the criteria in an explorative manner as they result from intensive analysis of the standard documents and literature. They were based on the main differences that we found during the analysis of the standards between life cycle assessment and the approach to carbon offsetting. The applied criteria and their guiding questions are displayed in Table 14.1.

Next, we *determined methodologies with the largest amount of issued credits with the goal to identify the most relevant ones.* We did this by accessing the programs' databases containing the issuance figures of the methodology types. The VCS's and CDM's databases and issuance figures are available online[2]; the Gold Standard's was provided by Claire Willers (Communications Manager of the Gold Standard) via email and was from the third quarter of 2018. We analyzed all methodologies covering more than 5% of total issued emission reductions for the respective program. These cover all relevant sectors in which carbon offsets are sold to date (industrial gas destruction, waste, renewable energy, forestry and energy efficiency). VCS and Gold Standard accept CDM standards when they comply with their own terms of requirements; therefore, the Gold Standard and VCS use the methodologies published by the CDM. We excluded the land use and forestry methodologies by the Gold Standard and the Climate Community and Biodiversity Standard for the VCS, because they include a biodiversity assessment that could not be evaluated in detail in this study. *Next, we allocated methodologies to sectors.* These sectors were oriented towards the

[2]https://www.vcsprojectdatabase.org/#/vcs (accessed January 2019).

https://cdm.unfccc.int/Projects/projsearch.html (accessed January 2019).

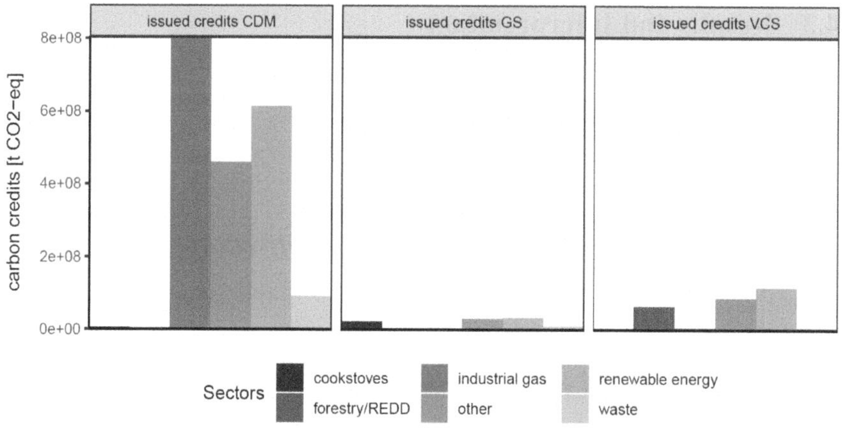

Fig. 14.2 Issued emission reductions for CDM, VCS and Gold Standard (GS)

so-called project types in the CDM. The results of step two and three are summarized in Fig. 14.2.

Subsequently, *we analyzed the sectors based on literature* (mainly the methodology documents, scientific publication, case studies, news coverage). The analysis was based on four criteria: considered life cycle phases, uncertainties, trade-offs and synergies, that have been identified as important aspects in step one. Our focus was to identify the considered life-cycle-phases of the methodologies. Moreover, we determined uncertainties related to challenges in emission accounting that are specific to a certain sector based on scientific literature and reports. Furthermore, we searched for case studies from literature to show which trade-offs regarding the environment and social aspects are relevant. For environmental trade-offs, the provided information from case studies were assigned to default impact categories in environmental LCA to show risks regarding possible trade-offs. Social issues were also considered—therefore, social impacts were assigned to subcategories from the Social Life Cycle Assessment (SLCA) Guideline (Althaus et al. 2009). We did this to display the trade-offs more systematically in Table 14.3. The result of this step is a qualitative analysis of each sector in text form (see Sect. 3.2).

Finally, we summarized *the results of step four regarding uncertainties, trade-offs and synergies as well as considered life-cycle-phase in* Table 14.3 with the objective to provide an overview of the life cycle perspective in the dominant carbon offsetting sectors.

14.3 Results and Interpretation

The results are structured in the generic analysis (3.1) and the sector specific analysis (3.2) as shown in Fig. 14.1.

14.3.1 *Comparison of ISO Standard with Offsetting Programs*

The results of the first step (comparison of ISO 14040/44 and ISO 14067 and the offsetting programs) are displayed in Table 14.2 and are listed in the following with further explanation:

1. Analyzed system: while LCA usually focusses on products or services and quantifies their function (ISO ISO 2006a, b, 2018), carbon offsetting methodologies always have a project focus.

Table 14.2 Comparison of LCA and carbon offset programs

Criteria	LCA/carbon footprint	Carbon offsets
1. Analyzed system	According to a functional unit	Project (no functional unit)
2. Scenario analysis	Possible	Mandatory
3. Life-cycle phases	Complete life-cycle, if not stated otherwise	Varying
4. Output	Information about potential environmental impacts of a product/process	Creates offsets (carbon credits) that have a value and leads to their production
5. Review	Critical review	Validation and certification by certification body
6. Data verification in review	Optional; exception PEF	Yes, detail depends on methodology
7. Flexibility in data collection and system boundaries	Data collection, system boundary and cut-offs are defined in goal and scope, but can be adapted in an iterative process during the study	Methodologies fix system boundaries and data quality requirements at the beginning of the project
8. Treatment of shift of burden and impact categories	Consideration of multi-impacts to avoid shift of burden; carbon footprint single impact, but makes no claim regarding other impacts	Single impact; to avoid shift of burden in foreground system Environmental Impact Assessment (EIA) is performed
9. Aggregation of biogenic and fossil carbon	Different sources that cause carbon emissions should be listed distinctly (according to ISO 14067)	No biogenic carbon accounting in CDM, but in VCS and GS; used to compensate for fossil emissions

2. Scenario analysis: An attributional LCA accounts for all environmental impacts along the life cycle of a product, but not their indirect consequences (Guinée et al. 2018). Some LCAs use scenarios to identify an environmentally favorable outcome e.g. different End-of-life (EoL) scenarios (Cherubini et al. 2009). However, these scenarios are not used to subtract one scenario from the other (and assigning the delta as a credit). Exactly this is mandatory in carbon offsets (Kollmuss et al. 2008). The tools provided by the CDM describe how the project developer should develop the "baseline scenario" and the "project scenario", while the surplus emissions from the baseline scenario are assignable as a credit. The underlying assumptions within these carbon offsetting scenarios involve considerable uncertainties, and existing studies show that often strategic modeling occurred to maximize carbon credits (e.g. Seyller et al. 2016; Wara 2008; Schneider et al. 2010). The impact of the defined reference scenario is discussed under the section "uncertainties" for the relevant sector in Sect. 3.2.
3. Life cycle phases: LCAs and carbon footprints based on the ISO standards normally consider all life-cycle phases, which is different in offsetting. The phases considered depend on the applied offsetting methodology, e.g. the waste methodology ACM0001 only accounts for avoided emissions in the EoL phase (UNFCCC, CDM 2017a), while the renewable energy methodologies only consider the use phase (UNFCCC, CDM 2018a). This aspect is discussed more detailed in a sector specific manner in Sect. 3.2.
4. Outputs: environmental offset programs create value i.e. the result of the assessment becomes a commodity; therefore, the project participants have an interest in a certain outcome (as Wara (2008) showed in relation to industrial gas projects and Seyller et al. (2016) in relation to forestry). The underlying problem is strongly related to 3: the reliability of the scenario analysis and is thus discussed in relation to the sectors' uncertainty in Sect. 3.2. The financial incentive to produce certain results is not so prevalent in LCA because it is often used for company internal information provision. As other fields of application are evolving for LCA (e.g. the product environmental footprint (PEF) aims to generate an ecolabel based on LCA (Communication from the Commission to the European Parliament and the Council—building the single market for green products 2013; European Commission 2018; Minkov et al. submitted)) prevalence of strategic behavior might increase.
5. Review: Only when an LCA case study is used for "comparative assertions", it must be reviewed by a review panel (ISO 2006), whereas many studies apply it voluntarily. Carbon offsets require validation and certification by an accredited certification body (including site visits). Existing studies (e.g. Schneider 2007; Seyller et al. 2016; Schneider and Möhr 2009) show that the certification is no barrier to unrealistic baseline scenarios.
6. Data verification and review: the critical review in LCA typically focusses on assumptions and chosen systems and the symmetry of the studies but not on data verification. Just the plausibility of the data with regard to the scope of the study is checked. Within PEF, which is based on LCA and the according ISO standards 14040/44 and aims to make LCA case studies more comparable, verification

and validation is required, including data verification (Communication from the Commission to the European Parliament and the Council—building the single market for green products 2013; European Commission 2018). However, the concept is not implemented yet. The verification and validation in the carbon offsetting programs requires the verification of the data, but the reliability of the results have been doubted (Schneider 2007; Schneider and Möhr 2009).

7. Flexibility in data collection and system boundaries: in LCA the scope is determined for each LCA in an iterative process so that the used data quality and system boundary can be determined to fulfill the study's goal (ISO 2006). Since LCA is such a diverse tool with different application fields, the flexibility in the ISO 14040 is necessary. For LCAs that are used in a product labelling context, predefined goal and scope definitions exist (including impact categories to choose and the data collection procedure), so called Product Category Rules according to ISO 14025 which often follow ISO 14040/44 requirements (Magerholm Fet and Skaar 2006). The standard for the carbon footprint (ISO 14067) presets the impact assessment method to use as well. The offsetting methodologies contain detailled guidance on how emissions should be accounted for, what the cut-offs are, and what assumptions can be made. They are therefore less flexible.
8. Treatment of shift of burden and impact categories: an LCA case study has several impact categories that are investigated to identify shifts of environmental burden. Example: bioplastics perform better in the category abiotic resource depletion compared to fossil fuel-based plastics, but have higher land and water use (Gironi and Piemonte 2011). Carbon offset methodologies account for carbon and do not aim at quantifying other environmental impacts, but just request an Environmental Impact Assessment if impacts are considered significant by the country where the project is developed (UNFCCC, CDM 2017b). The Environmental Impact Assessment is performed according to the legislative requirements of the country where the project is realized. E.g. India does not request an Environmental Impact Assessment for renewable energy projects, apart from large hydropower projects (Krithika and Mahajan 2014). Single impact 'LCAs' like carbon footprint or water footprint do not make statements about the effects caused in other impact categories.
9. Aggregation of biogenic and fossil carbon: ISO 14067—the standard for the carbon footprint—recommends to show the emissions of biogenic carbon, fossil carbon and carbon released for flying separately (ISO 2018). Mackey et al. argue against the aggregation of biogenic and fossil carbon in offsets (Mackey et al. 2013). A position that is also supported in the LCA discourse by Finkbeiner et al. (2013). Some carbon credit retailers do not follow this guidance especially since they sell credits to offset emissions from fossil sources with increased biogenic carbon stock or offer offsetting for the aviation sector.

14.3.2 Evaluate Relevant Sectors

Figure 14.2 shows the shares of the total carbon-credit issuance by sector for CDM, VCS and Gold Standard, respectively. All programs have a third of their credits generated by renewable energy projects. Only eight methodologies cover over 70% of issued CDM and VCS credits and just under 70% of Gold Standard credits and therefore we used them for the further investigation, while over 200 methodologies cover under one percent.

The eight most relevant methodologies were allocated to the following sectors: industrial gas (AM0001, AM0021), cookstoves (Improved cookstoves (ICs), AMS-II.G.), forestry and reduced emissions from deforestation and forest degradation (REDD) (VM0007, VM0009), renewable energy (ACM0002, AMS-I.D.) and waste (ACM0001).

The results are similar to a previous study from 2009, that found that the scope and methodologies in the voluntary and compliance market are similar (Corbera et al. 2009), but we also identified some changes. While renewable energy and landfill gas projects are used by all programs, industrial gas use is dominant in the CDM, while forestry is most relevant in VCS and not relevant for the CDM. Improved cookstoves only represent a dominant share of emission reductions for the Gold Standard.

In the following subchapters, each sector is analyzed regarding life-cycle emission accounting, challenges of calculations and uncertainties, possible trade-offs and synergies in the social and environmental dimension. Further, it is shortly addressed how this aspect is handled in LCA.

Forestry and REDD The VCS methodologies for REDD are inspired and based on the REDD + initiatives by the UNFCCC and were the first to introduce REDD in carbon compensation (Angelsen 2012). They provide guidance on the definition of project boundaries: geographical boundaries, reference scenarios and how to account for them. The methodology for avoided ecosystem conversion (VM0009) can only be applied if it is known that a certain piece of land will be converted (Verified Carbon Standard, Verra 2014). The project developers must identify drivers of conversion, which can be companies or subsistence farmers alike. The REDD methodology (VM0007) has a broader scope, that focusses also on illegal extraction of wood for fuel, but not for timber as well as planned and unplanned deforestation (Verified Carbon Standard, Verra 2015). In the subsequent paragraphs we will outline the considered life cycle phases, uncertainties, trade-offs and synergies.

- Life cycle emission accounting: Within LCA the entire life cycle of wood is taken into account and delayed emissions, like storage of CO_2 in products is not part of a carbon footprint according to ISO 14067, but they can be quantified and displayed in a separate assessment (ISO 2018). Since the forest is not a product in a strict sense, the forests life cycle is not mentioned in the methodologies. Only the avoidance of illegal wood extraction for fuelwood is included in the accountable carbon offset, but avoided wood extracted for timber cannot be accounted for (Verified Carbon Standard, Verra 2015). This leads to two challenges: (i) delayed emissions are allowed, and (ii) the full life cycle (like decomposition or incineration for the end of life) is not considered. When wood is extracted illegally from a forest it is very difficult to track for what purpose (timber or fuelwood) it is used, and therefore the scenario development might face high uncertainties.

In the following sources of uncertainty for forestry and REDD are discussed:

- The reference use-scenario is based on historical deforestation rates in the respective forest or in a forest close to the forest area that should be protected. If high baseline deforestation rates are assumed, the impact of REDD is likely to be overestimated (Seyller et al. 2016). The challenge of defining an appropriate reference scenario also applies to current LCIA methods for land use (Helin et al. 2013; Michelsen and Lindner 2015), but it is recommended to use natural relaxation (Koellner et al. 2013). In the carbon offset methodologies it is recommend to use historic deforestation rates, which are easier to manipulate (Seyller et al. 2016).
- Politically unstable countries: some countries in which REDD projects are implemented are politically unstable (Vijge et al. 2016). Political decision making or loss of control over the area will increase deforestation rates significantly (Seyller et al. 2016). Additionally weak enforcement of conservation or legislative amendments, but also increased economic development might significantly change historic deforestation rates (Dooley 2014). This aspect is also not considered within current LCIA methods for land use.
- Carbon content of landscapes: data can be taken from high-resolution digital imagery to calculate ex-ante forests carbon content, but this is accompanied with a high uncertainty (Dooley 2014). Furthermore, there is no consensus on the role of forests carbon stock related to increasing carbon concentration in the atmosphere and whether forests can be considered long term carbon sinks (Bellassen and Luyssaert 2014; Green et al. 2019; Reichstein et al. 2013). This aspect is currently not addressed in LCA.
- Longevity of the sequestered carbon and its limited capacity: Carbon uptake by land is not a permanent carbon sink, since carbon molecules just equilibrate between atmosphere, ocean and land.[3] Many forestry projects aim to monitor for 100 years. Mackey et al. (2013) suspect that this might be due to a misunderstanding of the 100-year time span chosen by the Intergovernmental panel on climate change (IPCC) (Forster et al. 2007) for Global Warming Potential. This is also a discussion in LCA and new approaches to overcome this asymmetry are under development (e.g., dynamic LCA (Levasseur et al. 2010; Levasseur et al. 2013)).

[3]Fossil carbon, that is extracted for incineration has been separated from the global carbon cycle for millions of years. Once it is emitted to the atmosphere by combustion, it is part of the global carbon cycle. It takes a shorter time (50–200 years) to adsorb a carbon molecule from the air to the ocean or the land in plants, that constitutes a new equilibrium, leaving a fraction of about 20% carbon in the atmosphere in the new equilibrium. However, these emissions result in an equilibrium with increased absolute carbon content between the lands' the atmospheres and oceans. The process that really removes the fossil carbon is not in a century range, but from 10,000 year and longer. All rephrased from Archer et al. (2009).

- Forest fires: the VCS considers unexpected reductions of the forest carbon stock by the creation of a buffer account. This buffer account should allow hedging of the respective credits, i.e. not the full amount of credits can be issued. The amount in the buffer account is cancelled if an ecosystem is degraded (e.g. due to illegal logging, drought or flood) or forest is lost due to fires. The VCS requires the monitoring of the projects for up to 100 years (Verified Carbon Standard, Verra 2017) with a buffer account of 10 to 40% of the issued credits (Seyller et al. 2016). These choices seem rather arbitrary considering for example the unknown uncertainty of the climate system's development and its influence on the forest carbon stock. Currently no better estimates for these challenges are available and also ISO 14067 notes that buffer and reserve accounts can mitigate the effect of non-permanence (ISO 2018). Unexpected reductions in the forest carbon stock like forest fires are not included in LCA, since LCA usually does not include risks in the inventory or the impact assessment.
- Forest/ecosystem capacity to take up carbon: even if all ecosystems, that were subject to land use change were restored this would reduce the CO_2 concentration only by 40–70 ppm, and even this unrealistic scenario cannot even out all fossil-based emissions (Mackey et al. 2013). The publication tree restoration potential in science came to a different conclusion, that forest could even out two third of historic emissions (Bastin et al. 2019) but has been criticized as well for this claim by climate experts like Rahmstorf (2019). This dispute shows that there is currently no scientific consensus.
- Social trade-offs: REDD and forestry projects kept indigenous and poor communities from entering forests (Sarmiento Barletti and Larson 2017). A case study in Nigeria reported militarization of forest protection in a government lead REDD + project (Asiyanbi 2016). Another case study from Madagascar that used a VCS methodology revealed that poor people depending on slash and burn agriculture were resettled in the course of a REDD project and did not received adequate compensation (Poudyal et al. 2018). A forestry project that also used a VCS methodology in Kenya had negative impacts on the local community's possibility to use slash and burn agriculture and produce charcoal (Chomba et al. 2016). We assign these findings to the SLCA subcategories access to resources, delocalization and migration and respect for indigenous rights.
- Synergies: if REDD projects ensure indigenous peoples' access to resources, do not prohibit local communities from entering forests and install biodiversity monitoring, the protection of forest ecosystem might benefit indigenous people (van Dam 2011) and biodiversity (Phelps et al. 2012) alike.

The analysis showed that the calculation of carbon credits from forestry contains significant challenges that are not solvable with the application of LCA instead. The knowledge about the forest, its carbon content and its role in the forest carbon cycle need to be better understood to model them more adequately. Further, the potential negative social impacts of the forestry projects are difficult to control in countries with poor governance. Moreover, forestry projects are completely independent from fossil fuel emissions that are the main cause of climate change.

Renewable Energy The renewable energy methodologies ACM0002 (UNFCCC, CDM 2018a) and AMS-I.D. (UNFCCC, CDM 2014) assess wind, hydro, geothermal and solar power. In the following, their life cycle emission accounting, uncertainties, trade-offs and synergies are presented:

- Life cycle emission accounting: the methodologies' scope is limited to the use phase of the renewable energy projects and assumes that project emissions for wind energy and solar energy are zero (UNFCCC, CDM 2014, 2018a). Upstream emissions from extraction, processing and transport (of utilized resources) are neglected (UNFCCC, CDM 2018a) or not even mentioned in AMS-I.D. Neither of the methodologies mention the EoL phase, in which the renewable energy technologies have to be recycled. Within LCA case studies, the entire life cycle is taken into account. LCA studies of renewable energy show that zero carbon emissions is neither a valid assumption for solar energy (Varun and Prakash 2009) nor for other renewable energy systems (Asdrubali et al. 2015).
- Uncertainties: the methodologies assume that energy produced by the project replaces energy that would otherwise be produced by the grid (UNFCCC, CDM 2014, 2018a) and, thus, mainly using fossil fuels, which most grids are dominated by. However, it is not always clear if fossil fuels will be replaced and how the introduction of renewable energy will affect the fossil fuel price (Foster et al. 2017). Usually an LCA study would only show the environmental impact of a certain technology, not its ability to replace another technology. The assumption that renewable energy capacity will directly replace fossil fuels capacity and thus suppress fossil fuel energy production has been proven wrong and the substitution effect was found to be lower than a quarter (York 2012). This finding challenges the key assumption of the carbon offset renewable energy methodology.

In the following the respective trade-offs are listed:

- Abiotic resource use: the impact of abiotic resource use is completely neglected in the offsetting methodologies, even though solar and wind energy depend partly on scarce resources like gallium, silicon, rare earth elements (REEs), indium and others (Viebahn et al. 2015). Several LCA based methods exist that consider resource depletion and scarcity (e.g. Schneider et al. 2011, 2016; Bach et al. 2016; van Oers and Guinée 2016).
- Toxicity: the sourcing of REEs and treatment in the EoL phase are associated with toxicity, e.g. REEs are often needed in permanent magnets in new model wind turbines (Marx et al. 2018). Another concern related to toxicity is the EoL phase of solar panels in which heavy metals like lead and cadmium can enter the groundwater if the solar cells are disposed of in landfills (Xu et al. 2018).
- Trade-offs of hydropower: Three large hydropower projects in Brazil lead to deforestation, displacement and significantly underestimate transboundary effects when damming the Amazon river (Yan 2012). Other hydropower projects involved resettlement without adequate compensation, indigenous right violation, destruction of cultural heritage and negative impacts on biodiversity (Obergassel et al.

2017). We assigned these impacts to the SLCA subcategories respect for indigenous rights and delocalization and migration and to biodiversity for environmental LCA. Deforestation and loss of farmland were assigned to land use change.
- Biodiversity and land use: A review by Gasparatos et al. showed that all renewable energy methods do interfere to some extend with biodiversity because of their land use that reduces available habitats (Gasparatos et al. 2017).
- Synergies: One opportunity—and co-benefit—of displacing fossil fuels by renewable energy is that other co-pollutants that occur in combustion processes, like NO_x and SO_x, that have an impact on acidification and human health (Smith et al. 2013). They can be significantly reduced by renewable energy. In the "offsetting" logic, these emissions could be offset in all energy-related projects that do not involve combustion, when the grid is dominated by fossil fuels. This is currently not done in carbon offsets.

Generally, many of the identified challenges and gaps related to the renewable energy methodologies can be tackled by applying LCA. The trade-offs could be identified, and all life cycle phases would be addressed. However, LCA could not solve the question to which extend renewable energy could displace fossil energy in the grid and what substitution factor would be valid.

Energy Efficiency Measures — Improved Cookstoves There are two relevant improved-cookstoves methodologies in the offsetting programs: the Gold Standard's simplified cookstoves methodology (The Gold Standard 2013) and the CDM methodology AMS-II.G. (UNFCCC, CDM 2018b) that can also be used in the VCS. These methodologies provide guidance how the fuel-efficiency of new and old stoves have to be measured and how the reduction in needed fuelwood has to be calculated. Life cycle emission accounting, uncertainties, and synergies are displayed in the following:

- Life cycle emission accounting: production and EoL of stoves are not taken into account (The Gold Standard 2013; UNFCCC, CDM 2018b). The reduced fuelwood demand and its net caloric value add up to the creditable emission reduction. In an LCA, all steps would have to be considered, which was done in an LCA study on an improved cookstoves project (Wilson et al. 2016).
- Uncertainties: for the baseline emission quantification AMS-II.G. assumes that saved energy would be produced by fossil fuels, which is very unlikely (UNFCCC, CDM 2018b). The Gold Standard methodology assumes that non-renewable biomass is used and therefore methane, black carbon (The Gold Standard 2017a) and nitrous oxide emissions are emitted during the incineration (The Gold Standard 2013). The assumption, that the firewood would be otherwise substituted by fossil fuels influences the results strongly, because otherwise they would not be eligible for the CDM. The CDM cannot include increased carbon stocks from reduced fuelwood consumption in its emission reductions due to the Marrakech Accord (Lee et al. 2014). Large uncertainties remain regarding parallel use of new cookstoves and old stoves, the amount of unrenewable biomass used, that is likely to be overestimated in the CDM default factors, and the determination

of the renewable/non-renewable biomass ratio (Lee et al. 2014). These problems were verified by an LCA study that showed that the use phase contains considerable uncertainty (Wilson et al. 2016). Khandelwal et al. (2017) showed that nearly none of the improved cookstoves are in use in India today, because they are more difficult to repair and the wood has to be cut.
- Inability to reduce fossil emissions: the improved cookstoves are mostly used with wood, not with fossil fuels. The improved cookstoves need less wood and thus reduce the incineration of wood, but not the burning of fossil fuels, which are the main driver of climate change. All uncertainties and risks that are associated to the increase of the forest carbon stock listed in the subsection of REDD and forestry also apply to the cookstoves methodologies.

Synergies:

- Human health: because of reduced combustion, reduced amounts of particulate matter are emitted, and thus indoor air quality is improved. This has a positive impact on human health (Khandelwal et al. 2017). The Gold Standard provides a tool to account for adverted disability adjusted life years due to improved cookstoves (The Gold Standard 2017b).
- Biodiversity and land use: reduced use of fuelwood can have positive impacts on biodiversity and land use. These positive impacts are particularly difficult to determine. Khandelwal et al. (2017) doubt that reduced personal wood demand does automatically lead to reduced logging.

Further research and the application of new technologies for data gathering could provide further insights on the sustained use of cookstoves (Ruiz-Mercado et al. 2011). The positive impact of improved cookstoves contain uncertainty, but we did not identify a significant trade-off in the environmental or social dimension. An application of LCA instead of the carbon offsetting methodologies could increase the robustness of the results, because e.g. black carbon accounting is not allowed in the carbon footprint according to ISO 14067 (ISO 2018). It should be noted, that the uncertainties in the use phase and all aspects that are related to the hypothetically increased forest carbon stock are not solvable with the application of LCA instead as explained in relation to forestry/REDD.

Industrial Gas The most important industrial gas methodologies are AM0001 (HFC-23 incineration from HFC-22 production) (UNFCCC, CDM 2011) and AM0021 (N_2O decomposition from adipic acid production) (UNFCCC, CDM 2009). The methodologies assess the obtainable emission reductions by a waste generation rate. The maximum reductions can be reached if the entire waste stream is destroyed. Inspired by the Matthew-principle, factories with large production capacity could reduce large amounts of HFC-22 and N_2O emissions and thus issue high amounts of CERs. Subsequently, the challenges in life cycle emission accounting, uncertainties and trade-offs are listed.

- Life cycle emission accounting: the methodology does not account for life-cycle emissions of HFC-22 nor adipic acid and their respective uses and further

processing into products. They also do not account for the emissions regarding the built infrastructure for incineration or decomposition and the necessary resources for the N_2O-depeletion catalyst. In an LCA these aspects would have to be included.
- Uncertainties: the industries' activity is likely to have changed because of expected revenues: Industrial gas offsets had low price efficiency. While the cost for the destruction of HFC-23 was 10 cents per one t CO_2-eq, the revenue for a permit was significantly higher (Wara 2008; Schneider et al. 2010). Moreover, the generated CERs produced a larger revenue than the final products (HFC-22 and adipic acid). Therefore, strategic behavior occurred i.e. facilities over reported their emissions or produced more industry gases than they could sell. The CDM adjusted its guidelines so that facilities that started production after 2004 cannot issue credits. Schneider et al. (2010) showed that significant carbon leakage occurred when CER-generating factories in non-Annex I countries took over production from non-CER-generation factories in Annex I countries, that had N_2O reduction technologies in place.

Trade-offs:

- Toxicity, human health and safe and healthy living conditions: industrial gas offsets subsidize an industry with a large impact on the environment, that pollutes water and has impacts on human health (Ghouri 2009).

Industrial gas technologies are a mere End-of-pipe technology and do not have a permanent future in offsetting since they are not allowed in any emission trading scheme anymore, and HCF-23 emissions will be faded out through the Montreal protocol since the Kigali amendment (United Nations Industrial Development Organization 2016) entered into force 2019 and regulates the long-term phase-out of HFC-23. In LCA, the implementation of an End-of-pipe technology usually does not create an offset. It would reduce the environmental impact associated with the product, but that reduction would not be attributable to some other factory or process. The LCA would also not attribute the emission reduction to the entity paying for the filter, which is done in carbon offsets.

Waste The methodology ACM0001 flaring or use of landfill gas provides guidance on the calculation of carbon credits in landfill technology (UNFCCC, CDM 2017a). These technologies are only creditable if the according treatment of waste is not required in the country by law. The avoided methane emissions and substitution of fossil energy through the energy won by landfill methane can be credited. It is possible to obtain credits when the landfill gas is flared for the avoided methane emission. Next the life cycle emission accounting, uncertainties as well as environmental trade-offs and synergies are listed and explained.

- Life cycle emission accounting: the methodology only accounts for the emissions caused and avoided at the landfill site, not for the construction of the site or its maintenance. The methodology includes the transport of landfill gas e.g. the

compression in trucks and distribution networks but excludes the needed infrastructure for the landfill gas capture and heat/electricity generation. In an LCA the EoL treatment is allocated to a product, but LCAs exist that only focus on EoL treatment strategies using scenario analysis combined with MFA (Cherubini et al. 2009). The study found landfilling with gas collection to be one of the least favorable EoL scenarios, compared to energy recovery and sorting combined with recycling.
- Uncertainties: the challenges associated with this methodology are similar to those regarding renewable energy, since it is unclear whether energy from landfill gas emissions replaces fossil fuel energy 1:1. There is no explicit analysis of landfill gas electricity generation, but York (2012) considers waste-to-energy processes in his study and found that they have no superior substitution rate of fossil energy compared to renewables. In LCA waste to energy processes are sometimes assumed to displace grid energy. Also PEF includes energy recovery from landfill, which can be assigned to the product as an energy credit (European Commission 2018). Damgaard et al. (2011) state that each landfill needs to be modelled individually if concrete numbers are needed, since every landfill has specific properties and is affected by its climate region. Landfill gas projects are difficult to manage since every dump has its own structure and thus an own waste generation rate, that is difficult to predict. Therefore most projects have a history of underperformance (GAIA 2011; Willumsen 2007).

Trade-offs:

- Human health, eutrophication and toxicity: an LCA case study showed that landfill gas collection can reduce greenhouse gas impacts and the acidification potential compared to an open dump, but show that all kinds of landfill have a significant impact on human- and eco-toxicity (Damgaard et al. 2011). Cherubini et al. (2009) also find that landfilling has eutrophication potential unlike other waste treatment strategies, like thermal treatment and sorting and recycling. However, they showed that landfill gas collection can significantly reduce GHG emissions and acidification potential compared to open dumps.
- Access to material resources: A case study from South Africa, reports that waste pickers were not allowed to visit the landfill site after a landfill gas collection project was implemented (GAIA 2011). Therefore, some landfill project might affect the access of the local community to material resources.

Landfills are a sink for valuable materials but also cause a burden on the environment. The disposal of valuable raw materials makes continual extraction necessary. The study by Cherubini et al. (2009) provided some first insights on how significant the offset potential of sorting and recycling can be compared to an open dump and landfill. Since energy generation by waste is assignable as a credit in some LCA approaches the difference between the offsetting methodology and the LCA approach are less distinct that for the other methodologies.

Table 14.3 displays an overall summary of all sectors with the respective criteria.

Table 14.3 Summary of sector specific evaluation

Sector	LC-Stage	Uncertainties	Trade-offs	Synergies
Forest (REDD) and ecosystem conservation	Not applicable	Forest carbon content, leakage if people are displaced, leakage for indirect land use change, time dimension (biogenic carbon), limited stock	Access to natural resources, delocalization and migration Respect for indigenous rights	Biodiversity[a], respect for indigenous rights[a] If applied correctly[a]
Renewable energy	Use phase	Substitution rate	Biodiversity, land use, abiotic resource depletion, delocalization and migration (hydropower), respect for indigenous rights (hydropower), toxicity (containing REEs, EoL solar)	Acidification (reduced SO_X and NO_X compared to fossil energy), human health
Energy efficiency (cook stoves)	Use phase	Use phase values uncertainty, black carbon accounting, renewable/non-renewable biomass ration (biogenic carbon)		Human health, particulate matter and black carbon, biodiversity, reduced land use
Industrial gases	Single process	Impartial monitoring, probable scenarios manipulation	Toxicity (leakage), human health, safe and healthy living conditions	
Landfill gas flaring/incineration	EoL	Methane generation rate, leakage of methane, efficiency of electricity generation, resources lost	Human health, toxicity, access to material resources, eutrophication	Acidification (reduced SO_X and NO_X compared to fossil energy)

[a]=if applied correctly

14.4 Discussion

In the following, we discuss the used method and its influence on the results, starting with the comparison of the different standard bodies. The scope of the comparison of carbon offsetting methodologies and LCA based on the ISO standard was limited, since today many practitioners use LCA in significantly different ways and new creative assessment tools are evolving, like scenario LCA that were not included in the analysis. Other carbon offsetting programs including the Land Use and Forestry part of the Gold Standard and the Climate, Community and Biodiversity Standard of the VCS were outside of the studies scope. Therefore, further analysis of existing programs and approaches are needed that might have an increased potential in combination with LCA.

Regarding the identification of relevant methodologies, the results of the issuance shares (Fig. 14.2) are strongly influenced by the sourced data. It is probable that double counting of issued credits occurred since some projects are registered under the Gold Standard and CDM simultaneously. Especially, determination of the issuance shares of Gold Standard credits proved to be difficult since the received issuance data contained missing information, and 10% of credits were not assignable to a methodology.

This study does not cover methodologies that account for under 5% of the issued emission reductions for the respective programs and therefore makes no statement regarding their applicability to LCA. Some methodologies that are only used to a small extent, but promise more favorable results for a combination with LCA, are energy efficiency measures that reduce grid-electricity use, recovery of secondary raw materials (since some LCA methods give environmental credits to recycled materials (Gala et al. 2015)), and CSS-technologies. These methodologies need further analysis, but market effects (e.g. fossil fuel price reactions to renewable energy subsidies) will remain a significant calculation challenge for secondary raw materials and energy efficiency.

The considered environmental impacts in Table 14.3 are not exhaustive but are only based on impacts that we deemed significant as the literature mentions them repeatedly. Moreover, the information on CDM projects is highly politicized. For this reason, the literature contained heterogeneous claims. While the CDM and UNFCCC argue that the CDM has been a great success (Spalding-Fecher et al. 2012; UNFCCC 2018), some scholars refer to the CDM scam and state that it failed to reduce any emissions (e.g. (Dabhi 2009; Lohmann 2005; Gilbertson 2017). To obtain all related tradeoffs, synergies and their order of magnitude, LCA case studies of offsetting projects that use the respective methodologies would be needed.

14.5 Conclusion

The analysis has shown that, currently, active carbon offsetting programs do not consider the whole life cycle. The different methodologies focus either on the use phase only for example in the case of renewable energy and energy efficiency methodologies, or an improved EoL treatment for example in the case of waste methodologies or on a modification of a single process for example in the case of industrial gas methodologies. All methodologies except cookstoves contained indication that negative environmental or social impacts can occur.

Further research that aims to improve the quality of environmental offsets could profit by implementing a life cycle and multi-impact perspective. Nevertheless, many challenges remain. Carbon credits issued by the carbon offsetting programs are commodities based on scenario analysis, influenced by market effects and future developments. They also do not consider social and environmental side effects systematically and homogenously. Their assessment is based on aspects that contain considerable uncertainties e.g. the substitution rate of alternative energy sources or the increase of the forest carbon stock for which currently applied LCA approaches do not have solutions. To avoid social trade-offs and increase acceptance detailed know-how of cultural and political background in the region where the projects should be realized is indispensable. A global perspective regarding potential environmental impacts that includes knowledge of environmental impacts of mining and local practices regarding the EoL phase are needed to model the realistic life cycle impact of a project.

References

Aldy JE, Stavins RN (2011) Using the market to address climate change: Insights from theory and experience. NBER working paper series, vol 17488, Cambridge, MA

Althaus H-J, Arena AP, Bouamrane M et al (2009) Guidelines for social life cycle assessment of products

Angelsen A (ed) (2012) Analysing REDD+: challenges and choices. CIFOR, Bogor

Archer D, Eby M, Brovkin V et al (2009) Atmospheric lifetime of fossil fuel carbon dioxide. Annu Rev Earth Planet Sci 37(1):117–134. https://doi.org/10.1146/annurev.earth.031208.100206

Asdrubali F, Baldinelli G, D'Alessandro F et al (2015) Life cycle assessment of electricity production from renewable energies: review and results harmonization. Renew Sustain Energy Rev 42:1113–1122. https://doi.org/10.1016/j.rser.2014.10.082

Asiyanbi AP (2016) A political ecology of REDD+: property rights, militarised protectionism, and carbonised exclusion in Cross River. Geoforum 77:146–156. https://doi.org/10.1016/j.geoforum.2016.10.016

Bach V, Berger M, Henßler M (2016) Messung von Ressourceneffizienz mit der ESSENZ-Methode: Integrierte Methode zur ganzheitlichen Bewertung

Bastin J-F, Finegold Y, Garcia C et al (2019) The global tree restoration potential. Science 365(6448):76–79. https://doi.org/10.1126/science.aax0848

Bellassen V, Luyssaert S (2014) Managing forests in uncertain times. Nature 12.02.2014(506 Issue 7487)

Böhm S (2009) Clean conscience mechanism: a case from Uruguay. In: Böhm S, Dabhi S (eds) upsetting the offset: the political economy of carbon markets. MayFly Books

Cames M, Harthan RO, Füssler J et al (2016) How additional is the clean development mechanism? Berlin

Cherubini F, Bargigli S, Ulgiati S (2009) Life cycle assessment (LCA) of waste management strategies: landfilling, sorting plant and incineration. Energy 34(12):2116–2123. https://doi.org/10.1016/j.energy.2008.08.023

Chomba S, Kariuki J, Lund JF et al (2016) Roots of inequity: how the implementation of REDD + reinforces past injustices. Land Use Policy 50:202–213. https://doi.org/10.1016/j.landusepol.2015.09.021

Communication from the Commission to the European Parliament and the Council—building the single market for green products (2013) facilitating better information on the environmental performance of products and organisations: COM(2013) 196 final

Corbera E, Estrada M, Brown K (2009) How do regulated and voluntary carbon-offset schemes compare? J Integ Environ Sci 6(1):25–50. https://doi.org/10.1080/15693430802703958

Dabhi S (2009) Where is climate justice in India's first CDM project? In: Böhm S, Dabhi S (eds) Upsetting the Offset: The Political Economy of Carbon Markets. MayFly Books

Damgaard A, Manfredi S, Merrild H et al (2011) LCA and economic evaluation of landfill leachate and gas technologies. Waste Manag 31(7):1532–1541. https://doi.org/10.1016/j.wasman.2011.02.027

Dooley K (2014) Misleading numbers: the case for separating land and fossil based carbon emissions

European Commission (2018) Product environmental footprint category rules guidance. ec.europa.eu/environment/eussd/smgp/pdf/PEFCR_guidance_v6.3.pdf. Accessed 09 Feb 2019

Finkbeiner M, Neugebauer S, Berger M (2013) Carbon footprint of recycled biogenic products: the challenge of modelling CO_2 removal credits. Int J Sustain Eng 6(1):66–73. https://doi.org/10.1080/19397038.2012.663414

Forster P, Ramaswamy V, Artaxo P, Berntsen T, Betts R, Fahey DW, Haywood J, Lean J, Lowe DC, Myhre G, Nganga J, Prinn R, Raga G, Schulz M, Van Dorland R (2007) Climate change 2007: the physical science basis; contribution of working group 1 to the fourth assessment report of the intergovernmental panel on climate change (IPCC). Cambridge University Press, Cambridge, MA

Foster E, Contestabile M, Blazquez J et al (2017) The unstudied barriers to widespread renewable energy deployment: fossil fuel price responses. Energy Policy 103:258–264. https://doi.org/10.1016/j.enpol.2016.12.050

GAIA (2011) Carbon trading prolongs environmental racism at Africa's largest municipal dump

Gala AB, Raugei M, Fullana-i-Palmer P (2015) Introducing a new method for calculating the environmental credits of end-of-life material recovery in attributional LCA. Int J Life Cycle Assess 20(5):645–654. https://doi.org/10.1007/s11367-015-0861-3

Gasparatos A, Doll CNH, Esteban M et al (2017) Renewable energy and biodiversity: implications for transitioning to a green economy. Renew Sustain Energy Rev 70:161–184. https://doi.org/10.1016/j.rser.2016.08.030

Ghouri N (2009) The great carbon credit con: why are we paying the third world to poison its environment? Daily Mail UK 2009

Gilbertson T (2017) Carbon pricing: a critical perspective for community resistance

Gironi F, Piemonte V (2011) Bioplastics and petroleum-based plastics: strengths and weaknesses. Energy sources, part a: recovery, utilization, and environmental effects 33(21):1949–1959. https://doi.org/10.1080/15567030903436830

Green JK, Seneviratne SI, Berg AM et al (2019) Large influence of soil moisture on long-term terrestrial carbon uptake. Nature 565(7740):476–479. https://doi.org/10.1038/s41586-018-0848-x

Guinée JB, Cucurachi S, Henriksson PJG et al (2018) Digesting the alphabet soup of LCA. Int J Life Cycle Assess 23(7):1507–1511. https://doi.org/10.1007/s11367-018-1478-0

Hamrick K, Gallant M (2017) Unlocking potential: state of the voluntary carbon markets 2017, Washington, DC

Helin T, Sokka L, Soimakallio S et al (2013) Approaches for inclusion of forest carbon cycle in life cycle assessment—a review. GCB Bioenergy 5(5):475–486. https://doi.org/10.1111/gcbb.12016

Hellweg S, Milà i Canals L (2014) Emerging approaches, challenges and opportunities in life cycle assessment. Science (6188):1109–1113. https://doi.org/10.1073/pnas.1220362110

ISO (2006a) ISO 14044:2006 Environmental management—life cycle assessment—requirements and guidelines

ISO (2006b) ISO 14040:2006 Environmental management—life cycle assessment—principles and framework (ISO 14040:2006) (14040)

ISO (2018) ISO 14067:2018 Greenhouse gases—Carbon footprint of products—requirements and guidelines for quantification

Khandelwal M, Hill ME, Greenough P et al (2017) Why have improved cook-stove initiatives in India failed? World Dev 92:13–27. https://doi.org/10.1016/j.worlddev.2016.11.006

Koellner T, Baan Ld, Beck T et al (2013) UNEP-SETAC guideline on global land use impact assessment on biodiversity and ecosystem services in LCA. Int J LCA

Kollmuss A, Zink H, Polycarp C (2008) Making sense of the voluntary carbon market: a comparison of carbon offset standards

Krithika PR, Mahajan S (2014) Governance of renewable energy in India Issues and challenges, New Delhi

Lee CM, Chandler C, Lazarus M et al (2014) Assessing the climate impacts of cookstove projects: issues in emissions accounting. Chall Sustainab 1(2). https://doi.org/10.12924/cis2013.01020053

Levasseur A, Lesage P, Margni M et al (2010) Considering time in LCA: dynamic LCA and its application to global warming impact assessments. Environ Sci Technol 44(8):3169–3174. https://doi.org/10.1021/es9030003

Levasseur A, Lesage P, Margni M et al (2013) Biogenic carbon and temporary storage addressed with dynamic life cycle assessment. J Ind Ecol 17(1):117–128. https://doi.org/10.1111/j.1530-9290.2012.00503.x

Lohmann L (2005) Marketing and making carbon dumps: commodification, calculation and counterfactuals in climate change mitigation. Science as Culture 14(3):203–235. https://doi.org/10.1080/09505430500216783

Mackey B, Prentice IC, Steffen W et al (2013) Untangling the confusion around land carbon science and climate change mitigation policy. Nature Clim Change (Nature Climate Change) 3(6):552–557. https://doi.org/10.1038/nclimate1804

Magerholm Fet A, Skaar C (2006) Eco-labeling, product category rules and certification procedures based on ISO 14025 requirements. Int J Life Cycle Assess (11)

Martínez-Blanco J, Inaba A, Finkbeiner M (2015) Scoping organizational LCA—challenges and solutions. Int J Life Cycle Assess 20(6):829–841. https://doi.org/10.1007/s11367-015-0883-x

Marx J, Schreiber A, Zapp P et al (2018) Comparative life cycle assessment of NdFeB permanent magnet production from different rare Earth deposits. ACS Sustain Chem Eng 6(5):5858–5867. https://doi.org/10.1021/acssuschemeng.7b04165

Masson-Delmotte V (2018) Global warming of 1.5 °C: an IPCC special report on the impacts of global warming of 1.5 °C. https://www.ipcc.ch/sr15/

Michelsen O, Lindner J (2015) Why include impacts on biodiversity from land use in LCIA and How to select useful indicators? Sustainability 7(5):6278–6302. https://doi.org/10.3390/su7056278

Obergassel W, Peterson L, Mersmann F et al (2017) Human rights and the clean development mechanism: lessons learned from three case studies. JHRE 8(1):51–71. https://doi.org/10.4337/jhre.2017.01.03

Pearce D (2002) An intellectual history of environmental economics. Annu Rev Energy Environ 27(1):57–81. 74p. https://doi.org/10.1146/annurev.energy.27.122001.083429

Phelps J, Friess DA, Webb EL (2012) Win–win REDD + approaches belie carbon–biodiversity trade-offs. Biol Cons 154:53–60. https://doi.org/10.1016/j.biocon.2011.12.031

Poudyal M, Jones JPG, Rakotonarivo OS et al (2018) Who bears the cost of forest conservation? PeerJ 6:e5106. https://doi.org/10.7717/peerj.5106

Rahmstorf S (2019) Can planting trees save our climate? http://www.realclimate.org/index.php/archives/2019/07/can-planting-trees-save-our-climate/. Accessed 10 Sep 2019

Reichstein M, Bahn M, Ciais P et al (2013) Climate extremes and the carbon cycle. Nature 500(7462):287–295. https://doi.org/10.1038/nature12350

Ruiz-Mercado I, Masera O, Zamora H et al (2011) Adoption and sustained use of improved cookstoves. Energy Policy 39(12):7557–7566. https://doi.org/10.1016/j.enpol.2011.03.028

Sarmiento Barletti JP, Larson AM (2017) Rights abuse allegations in the context of REDD + readiness and implementation: a preliminary review and proposal for moving forward. https://www.cifor.org/library/6630/. Accessed 06 Mar 2019

Schneider L (2007) Is the CDM fulfilling its environmental and sustainable development objectives?. An evaluation of the CDM and options for improvement, Berlin

Schneider L, Möhr L (2009) A rating of designated operational entities (DOEs) accredited under the clean development mechanism (CDM): scope, methodology and results, vol 2009

Schneider L, Lazarus M, Kollmuss A (2010) Industrial N_2O projects under the CDM: adipic acid—a case of carbon leakage? SEI working paper

Schneider L, Berger M, Finkbeiner M (2011) The anthropogenic stock extended abiotic depletion potential (AADP) as a new parameterisation to model the depletion of abiotic resources. Int J Life Cycle Assess 16(9):929–936. https://doi.org/10.1007/s11367-011-0313-7

Schneider L, Bach V, Finkbeiner M (2016) LCA perspectives for resource efficiency assessment. In: Finkbeiner M (ed) Special types of life cycle assessment. Springer Netherlands, Dordrecht, s.l

Seyller C, Desbureaux S, Ongolo S et al (2016) The "irtual economy" of REDD + projects: does private certification of REDD + projects ensure their environmental integrity? Int Forest Rev 18(2):231–246. https://doi.org/10.1505/146554816818966273

Smith KR, Frumkin H, Balakrishnan K et al (2013) Energy and human health. Annu Rev Public Health 34:159–188. https://doi.org/10.1146/annurev-publhealth-031912-114404

Spalding-Fecher R, Achanta AN, Erickson P et al (2012) Assessing the Impact of the clean development mechanism

The Gold Standard (2013) Methodology for improved cook-stoves and kitchen regimes V.01: indicative programme, baseline, and monitoring methodology for improved cook-stoves and kitchen regimes

The Gold Standard (2017a) Quantification of climate related emission reductions of black carbon and co-emitted Species due to the replacement of less efficient cookstoves with improved efficiency cookstoves version 1.0. https://globalgoals.goldstandard.org/401-13-cr-slcp-gold-standard-quantification-of-climate-related-emission-reductions-of-black-carbon-and-co-emitted-species-due-to-the-replacement-of-less-efficient-cookstoves-with-improved-efficiency/. Accessed 15 Feb 2019

The Gold Standard (2017b) Methodology to estimate and verify averted mortality and disability adjusted life years (ADALYs) from cleaner household Ai. https://globalgoals.goldstandard.org/401-3-adalys-from-cleaner-household-air/. Accessed 20 Mar 2019

The Gold Standard (2019) Gold Standard CDM FAQ

UNFCCC (2018) Clean development mechanism passes new milestone. https://unfccc.int/news/clean-development-mechanism-passes-new-milestone. Accessed 19 Mar 2019

UNFCCC, CDM (2009) Approved baseline and monitoring methodology AM0021 Version 03: Baseline Methodology for decomposition of N_2O from existing adipic acid production plants. http://cdm.unfccc.int/filestorage/C/D/M/CDMWF_AM_BCEJNPA3ZPOWZOE318KULMRZPBLLWF/EB45_repan06_AM0021_ver03.pdf?t=dzF8cG15dXBofDBWkUxUvpEsfCPksfXQvw5n. Accessed 15 Feb 2019

UNFCCC, CDM (2011) Approved baseline and monitoring methodology AM0001 Version 06.0.0: decomposition of fluoroform (HFC-23) waste streams. http://cdm.unfccc.int/filestorage/5/0/K/50KH2J9V6O1IQNBSPALXYUGRCZFED7.1/EB65_repan10_AM0001_ver06.0.0_v02.pdf?t=cnp8cG15dWtpfDAmm7rFU8rr2Hrd0QIREBQr. Accessed 15 Feb 2019
UNFCCC, CDM (2014) AMS-I.D. Small-scale Methodology Version 18.0: Grid connected renewable electricity generation. https://cdm.unfccc.int/filestorage/2/P/7/2P7FS6ZQAR84LG3NMKYUH50WI9ODBC/EB81_repan24_AMS-I.D_ver18.pdf?t=VE98cG15dWYzfDBIhuxjR6LryACgpf2gZwJW. Accessed 15 Feb 2019
UNFCCC, CDM (2017a) ACM0001 Large-scale consolidated methodology version 18.1: flaring or use of landfill gas. https://cdm.unfccc.int/methodologies/DB/Y88077XT5O83TZ2PYEZ36LFIAMAODR. Accessed 15 Feb 2019
UNFCCC, CDM (2017b) Standard CDM project standard for project activities: Version 02.0. https://cdm.unfccc.int/filestorage/e/x/t/extfile-20181221092046529-Reg_stan04v02.pdf/Reg_stan04v02.pdf?t=Z2V8cHVmNDVufDByN_e0jTRvVfD5IY_hcBgL. Accessed 10 Jul 2019
UNFCCC, CDM (2018a) ACM0002 Large-scale consolidated methodology version 19.0: grid-connected electricity generation from renewable sources. https://cdm.unfccc.int/filestorage/5/8/I/58IAGB7SZUDEO2VN6LYM30K41HFPRQ/EB100_repan06_ACM0002.pdf?t=ZlR8cG15dWF0fDB9lvs9KON_B50wTgkYufW7. Accessed 15 Feb 2019
UNFCCC, CDM (2018b) AMS-II.G. Energy efficiency measures in thermal applications of non-renewable biomass Version 10. https://cdm.unfccc.int/filestorage/1/F/S/1FSPVQM7JWELKHB5U94DXR23TOC6AZ/EB100_repan12_AMS-II.G.pdf?t=bTR8cG55YnJofDCyJ1dX1GMhNBopIXZYnEqo. Accessed 06 Mar 2019
UNFCCC, CDM (2019a) CDM methodologies. https://cdm.unfccc.int/methodologies/index.html. Accessed 26 Feb 2019
UNFCCC, CDM (2019b) CDM project cycle. http://cdm.unfccc.int/Projects/diagram.html. Accessed 26 Feb 2019
United Nations Industrial Development Organization (2016) The Montreal protocol evolves to fight climate change. https://www.unido.org/sites/default/files/2017-07/UNIDO_leaflet_07_MontrealProtocolEvolves_170126_0.pdf. Accessed 10 Sep 2019
van Dam C (2011) Indigenous territories and REDD in Latin America: opportunity or threat? Forests 2(1):394–414. https://doi.org/10.3390/f2010394
van Oers L, Guinée J (2016) The abiotic depletion potential: background, updates, and future. Resources 5(1):16. https://doi.org/10.3390/resources5010016
Varun Bhat IK, Prakash R (2009) LCA of renewable energy for electricity generation systems—a review. Renew Sustain Energy Rev 13(5):1067–1073. https://doi.org/10.1016/j.rser.2008.08.004
Verified Carbon Standard, Verra (2014) Approved VCS methodology VM0009 version 3.0: methodology for avoided ecosystem conversion. https://verra.org/wp-content/uploads/2018/03/VM0009-Methodology-for-Avoided-Ecosystem-Conversion-v3.0.pdf. Accessed 15 Feb 2019
Verified Carbon Standard, Verra (2015) VCS methodology VM0007, Version 1.5: REDD + methodology framework (REDD-MF). https://verra.org/wp-content/uploads/2017/10/VM0007v1.5.pdf. Accessed 15 Feb 2019
Verified Carbon Standard, Verra (2017) Agriculture, Forestry and Other Land Use (AFOLU) Requirements v3.6. http://verra.org/wp-content/uploads/2018/03/AFOLU_Requirements_v3.6.pdf. Accessed 16 Feb 2019
Viebahn P, Soukup O, Samadi S et al (2015) Assessing the need for critical minerals to shift the German energy system towards a high proportion of renewables. Renew Sustain Energy Rev 49:655–671. https://doi.org/10.1016/j.rser.2015.04.070
Vijge MJ, Brockhaus M, Di Gregorio M et al (2016) Framing national REDD + benefits, monitoring, governance and finance: a comparative analysis of seven countries. Glob Environ Change 39:57–68. https://doi.org/10.1016/j.gloenvcha.2016.04.002
Wara M (2008) Measuring the clean development mechanism" performance and potential. Article in UCLA Law Review, University of California

Willumsen H (2007) CDM—landfill gas projects: introduction to discussion World Bank Workshop, Washington DC

Wilson DL, Talancon DR, Winslow RL et al (2016) Avoided emissions of a fuel-efficient biomass cookstove dwarf embodied emissions. Dev Eng 1:45–52. https://doi.org/10.1016/j.deveng.2016.01.001

Xu Y, Li J, Tan Q et al (2018) Global status of recycling waste solar panels: a review. Waste Manag 75:450–458. https://doi.org/10.1016/j.wasman.2018.01.036

Minkov N, Annekatrin L, Finkbeiner M The product environmental footprint communication at the crossroad: integration into or co-existence with the European Ecolabel? Int J LCA (submitted)

Yamin F (ed) (2014) Climate change and carbon markets: a handbook of emission reduction mechanisms, paperback. Earthscan, London

Yan K (2012) Carbon offsets misused by hydropower industry. World River Rev (June)

York R (2012) Do alternative energy sources displace fossil fuels? Nat Clim Change 2(6):441–443. https://doi.org/10.1038/nclimate1451

Chapter 15
Comparability of LCAs — Review and Discussion of the Application Purpose

Maximilian Roßmann, Matthias Stratmann, Nadine Rötzer, Philipp Schäfer, and Mario Schmidt

Abstract This article discusses the comparability of Life Cycle Assessments (LCAs) and the central role of the application purpose in a study review. According to ISO 14040, an LCA study design emerges in continuous reference to the "intended application". Goal and scope, case-specific assumptions, as well as methodological freedoms, should be justified by their significance for the specific application purpose, e.g. for process optimization or for advice on a political issue. In contrast, our systematic review of 58 LCA studies shows that LCAs hardly name applications, and more generally, applications are difficult to reconstruct. This lack of transparency makes the LCA methodology attackable through meta-studies that ignore the problem-oriented and case-specific approach. Since these studies are valuated for different purposes by a diverse set of actors, quantification in any study that does not represent the context and purpose of its generation can disguise as much as it can enlighten. Therefore, we propose what a study should look like that is problem-solving, concrete and yet provides transferable results for other studies.

Keyword Life cycle assessment · Application · Comparability · Biofuels · Technology assessment · Systems theory

M. Roßmann (✉)
Institute for Technology Assessment and Systems Analysis (ITAS), Karlsruhe Institute of Technology, Karlstraße 11, 76133 Karlsruhe, Germany
e-mail: maximilian.rossmann@kit.edu

M. Stratmann · N. Rötzer · P. Schäfer · M. Schmidt
Institute for Industrial Ecology (INEC), Pforzheim University, Tiefenbronner Straße 65, 75175 Pforzheim, Germany

M. Schmidt
Faculty of Sustainability, Leuphana University Lüneburg, Universitätsallee 1, 21335 Lüneburg, Germany

S. Albrecht et al. (eds.), *Progress in Life Cycle Assessment 2019*, Sustainable Production, Life Cycle Engineering and Management, https://doi.org/10.1007/978-3-030-50519-6_15

15.1 Introduction

Life Cycle Assessments (LCAs) quantify the environmental impacts of products and services over the entire life cycle. They, therefore, play an important role e.g. in strategic planning, advertising and current social and political debates, such as the energy transition or world food supply. Thus, the European Commission (European Commission 2003) concluded that "life cycle assessments provide the best framework for assessing the potential environmental impacts of products currently available". Several ISO standards and several manuals, textbooks and regulations are meant to ensure and improve comparability, resilience as well as reproducibility within a given framework (European Commission—JRC 2010; Klöpffer and Grahl 2012; Hauschild et al. 2018; DIN EN 1404 2009). However, various meta-studies show how different interpretations and methodological choices within the given framework lead to very different results, as for example for packaging systems (von Falkenstein et al. 2010), aluminum applications (Liu and Müller 2012), biofuels (Martin et al. 2015; van der Voet et al. 2010) and the metal and mining sector (Yellishetty et al. 2009). One could say that the methodology is not approaching consensus, but seems to be diversifying (Klöpffer and Grahl 2012). The emerging "LCA-Spin-Offs", like Footprints or beyond-product-LCAs (OLCA, IO-LCA), nowadays, rather have the character of an "LCA alphabet soup" than of a uniform method with similar or robust assumptions (Guinée et al. 2018; Finkbeiner 2014). If an LCA is considered a basis of the advertisement or organizational strategy in a highly debated field, such opacity can lead to protracted disputes. For the purpose of evidence-based policy, modeling such as the LCA, in general, is even considered to be in a crisis (Saltelli and Giampietro 2017).

The mentioned ISO standards already stipulate that the preparation of LCAs is to be geared to their intended application. In the ISO standards, the purpose of application plays a central role; every step in the preparation and interpretation phase of a study should reflect the purpose (DIN EN 1404 2009). According to the standards, the definition of goal and scope, the selection of allocation methods and the assessment data quality should be oriented towards the intended purpose. Facing the differences between the studies, this claim is already being discussed in the LCA community, for example in the distinction between policy LCAs, which should be particularly robust, and analysis LCAs, which aim to understand systems (Wardenaar et al. 2012). While Tjerk Wardenaar et al. (2012) consider the influence of the application on the allocation method, Christian Bauer, Liselotte Schebek, and Mario Schmidt (Bauer et al. 2007) see the influence on all levels of modeling, respectively the selection of modeled processes, flows and impact categories.

In this paper, we take up the debate and argue for inevitable methodological diversity and different assumptions to address particular problems of an application purpose. On the one hand, there are good reasons to claim for a stricter methodology, a general data basis for all further studies and the purist restriction to attributional LCAs (Guinée et al. 2018). On the other hand, model assumptions and methodological diversity are reasonable when the model is only measured by the purpose for which it is used (Saltelli 2019). The different demands of recognized plurality and a scientific-political consensus on the "correct LCA" are not contradictory but emphasize different perspectives. Both serve to reject the general critique regarding their opaque assumptions and unnecessary diversification by emphasizing the purpose of the study. We, therefore, question if an analysis of different application purposes can reject the claim for a stricter methodology and give perspectives for better LCA studies?

In a first step, we will discuss the application from a system-theoretical perspective and narrow down the term "application case" in order to make it empirically classifiable. In an analysis of 58 studies on biofuels, we show how little one can learn about the application purpose and justification of the model assumptions. In our discussion about the state and transferability of this critique, we will follow up on the debates about application purposes against the background of methodical freedom. We conclude that the pure reference to the correct application of the ISO or otherwise specified LCA standard does not justify a comparable, transparent decision basis for responsible policy processes and organizational planning. The LCA community, therefore, should take the critique seriously as a demand to better take into account and communicate the context and purpose of LCA studies.

15.2 Review of the Application Purpose

The argument to evaluate models according to their usefulness for a particular application is not limited to material flow models. We call different material and abstract objects models if they are used to represent a phenomenon of the world without matching the target in every aspect. Models represent, for example, the general balances or certain causal relationships of a target system, often involving generalized system knowledge. A model that matches in every aspect would be a duplicate of the target system. When modelers try to represent too many aspects, the additional assumptions create additional uncertainties that, taken together, reduce the scope and usefulness of the model (Saltelli 2019). For our consideration, it is only important that the model matches in so far that it is suitable for a certain purpose. The goal and scope of an LCA are to define the framework in which the model should represent the target system in order to draw attention to the characteristics of this representation. In science, models serve to learn about a target system. We learn with models, by constructing models and by varying parameters (Frigg and Nguyen 2016). The scientific communication about models serves to share procedures for generating certain experiences that are discussed and acknowledged as facts in the community. Within scientific communication, it is clear that truth statements refer to models and do not correspond to an external world: One the one hand, something is not considered objectively true, but only true according to this or that model (Walton 1990; Toon 2014). On the other hand, this or that model is discussed and considered suitable for a certain purpose, for example, to construct a factory building, to analyze energy costs, or to balance environmental emissions of a product lifetime. These aspects describe the goal and scope of a model. However, the intended purpose, and thereby the necessary abstraction and representation of complex contexts, refers to the social function of a model. How will people use the LCA to make a difference?

If one follows Niklas Luhmann's system-theoretical distinctions, models in politics serve to gain power and to promote collectively binding decisions (Luhmann 1986, 1987). That means models are used to persuade others to recognize one's own assumptions as the basis for further consideration, action, and decision-making. But even for this, they must seek acceptance of shared bases. Scientifically authorized modeling shows ways for formulating truth claims that are considered true to the extent that their conclusion based on the assumptions can be understood and reproduced by everyone in a certain scientific community (Luhmann 1992). Since scientific truth is widely acknowledged in our society, scientific truth is linked to power. But the politically relevant conclusions often exceed the scientifically supported ones. Therefore, for example, further model assumptions must be collectively supported, for example by stakeholder participation. The normative conclusions, i.e. which practical measures and decisions stakeholders derive from modeling, must also be shaped politically. This challenge of dealing with normative assumptions in scientific policy advice has been the subject of ongoing critical and fruitful debates on technology assessment since the 1970s (Nierling and Torgersen 2019; Grunwald 2019; Wynne 1975). Besides the science-politics-interface, there are also perspectives on models from law and economics. At the point of institutional decisions, model assumptions and scientific opinions are woven into legal norm systems. This means that models that follow norms and hermeneutic blanks provide a general basis and interface on which particular cases can be interpreted and decided. Relying on legal standards ensures that assumptions do not have to be negotiated every time, but that all those possibly affected would be given justice in the same way. However, from an economic perspective, models, in the end, aim to make a profit. They can be used to convince others to invest or buy products (advertising and motivation), to show potential benefits and savings (process analysis), to position oneself strategically in the long term (orientation), or to prove the fulfillment of legal standards for e.g. tax benefits or funding. In order to perform these functions, models do not have to be supported by all those potentially affected, they do not have to be transparent in every respect and they do not have to be based on legal standards. But for different purposes, they can attempt that.

This systems theoretical perspective illustrates the environments in which modeling takes place: Even if the modeler does not follow any further intentions when creating his model, but only his intuition, his work can be observed and utilized in various perspectives. The functional system reference does not refer to a person, but to the communicative connection of his or her communication. A scientist's statements can, for example, also have political, legal and economic effects. On the other hand, we see that there are studies that would not satisfy scientific requirements but which form the basis for political and economic decisions, we see that there are scientific models that do not contribute to making a case legally decidable, and we see that there are industrial studies that elude scientific assessment because, for example, the data are not publicly accessible. As models have different application purposes, different characteristics are in the foreground. This general view of models in society can also be applied to LCAs. At first glance, it becomes clear that modeling anticipates different success criteria for different purposes without allowing a real prediction of practical success. In the next step, we observe degrees of freedom in the LCA, which can be aligned to a specific application purpose.

15.3 Assumptions and Methodological Freedoms

The common reference to ISO standards primarily shows that a legal framework is considered at the basis for modeling and assessment. The training for dealing with ISO standards in LCAs enables the normal interpretation of standards, so that in LCA practice, for example, a reviewer can speak of a correct or incorrect application of standards. Before the standardization and establishment of textbooks or technical journals, however, there were life cycle assessments, in particular, due to the interests of the industry. Matthias Finkbeiner calls these early times the "wild west times", in which the use of life cycle assessments was detrimental to credibility due to strong bias and misuse (Finkbeiner 2014). Nowadays, one speaks of an established method and a worldwide community, and many studies are divided into a goal and scope phase, an inventory phase, an impact assessment phase and an interpretation phase — just as recommended in the standards.

There are various references in the standards that the intended application belongs to the study and should be taken into account:

- "The goal and scope of an LCA shall be clearly defined and shall be consistent with the intended application" (Chap. 4.2.1 in (DIN EN 1404 2018)).
- "The choice of elements of the physical system to be modeled depends on […] its intended application and audience […]" (Chap. 5.2.3 in (DIN EN 1404 2009)).
- "Recommendations should relate to the intended application" (Chap. 4.5.4 in (DIN EN 1404 2018)).
- "Therefore, special care is necessary to ensure that the information is applicable to the context in which it is likely to be applied" (Annex A.2 in (DIN EN 1404 2009)).
- "Clarifications, considerations, practices, simplifications, and options for the different applications are […] beyond the scope of this International Standard" (Annex A.1 in (DIN EN 1404 2009)).

The latter point, in particular, indicates that a study has a purpose that is not specified by ISO. An LCA can have different purposes. However, it is not specified how these purposes are to be fulfilled. The modeler must decide for himself what is sufficient, essential or suitable. The same is true for freedoms shown by terms such as "suitable", "sufficient", "essential", "potential" and "relevant". Grunwald (2016) calls these terms in comparison with indefinite legal terms "hermeneutic blanks" as compared to indefinite legal terms, which must be filled with meaning for the particular case. An indefinite legal term is a term that is not filled in by a clearly defined fact but must be specified during the application of the law for every individual case. Further freedoms are shown above all in the choice of allocation, the attribution of environmental impacts to the different valuable outputs of a system (Wardenaar et al. 2012). Just as social coexistence is guided by formalized legal and social norms but cannot be determined, the LCA practice of formal standards and informal norms requires interpretation and good scientific practice. The attempt to take into account any special case would make the body of rules grow immeasurably. In this way, it has also been historically shown that science embraces versatile practices without a general formula or method (e.g. Lakatos and Musgrave 1974).

The practices of LCA seem too different for implicitly shared norms to emerge as a common ground for all application purposes. In their review, Wardenaar et al. (2012) find such large variations that they propose to distinguish generally between political LCAs aimed at the robustness and more diverse analytical LCAs for different more explorative purposes. On the one hand, the European Commission is developing the Product Environmental Footprint, which is a very standardized LCA-method, where "comparability is given priority over flexibility" (European Commission 2013). On the other hand, industrial users of LCAs try to include real process data if possible to optimize a process in a company. In the same way, it would be scientifically nonsense if standards were to prevent systems from being modeled and explored more appropriately on the basis of empirical data. In the cases of strategic consulting and foresight, LCA models are coupled with upscaling, market estimates and other model extensions already. However, from the perspective of policy-assessment, there is a growing concern about an increasing number of "similar-but-different" methodologies and approaches (Galatola and Pant 2014). Likewise, companies are more interested in satisfying customers' requests for information about environmental impacts by means of self-designed labels than in provoking and going through complex certification procedures. In summary, there are different applications and interests in conducting life cycle analysis, which involve different assumptions and promote model diversification. But can these purposes also be revealed in published studies?

15.4 Review of the Meta-study

In order to scrutinize the general criticism of LCAs, we analyzed the LCA studies considered in the meta-study by van der Voet et al. (2010) in terms of their intended application and assumptions made. In total, we were able to review 58 studies. The initial idea was to investigate if there is a relationship between application purpose and the different attributes (application, functional unit, system boundary, allocation method). However, this turned out to be not feasible as we will explain later. We, therefore, included further attributes such as the ranking of the journals (impact factor) and the reference to the ISO standard to explain differences between the studies.

In order to classify the application purpose of an LCA-study published in a paper, it is not sufficient to specify a set of search terms for automated text analysis. The reference to an application purpose does not seem to be sufficiently specified. The analysis thus consisted of a careful and iterative elaboration of the purpose of the application in the introduction and discussion and final section of the studies as common in discourse analysis. To give an example, the decision is briefly played through on a paper named "Can ethanol alone meet California's low carbon fuel standard? An evaluation of feedstock and conversion alternatives" (Zhang et al. 2010). On the one hand, the paper does not name an intended application purpose. On the other hand, the title refers directly to a specific application and the paper contains concrete references to the Low Carbon Fuel Standard (LCFS) program in California. It contains bits of advice like: "The inclusion of metrics other than solely GHG [Green House Gas] emissions offers insights potentially relevant for avoiding unintended consequences" (Zhang et al. 2010). Without wanting to deny the advertising effect and the scientific sharpness, this indicates the purpose of political consultation. The political intention is also represented in the subsequent acknowledgments to General Motors (GM) for sponsoring the study. Without now discussing the neutrality or standards of independence, it can be postulated that this study is only politically successful if it makes assumptions that are robust enough not to be easily refuted by political opponents. The study thus carries the political story that in signing the law it is better to take certain factors into account—not in order to do GM a favor but because otherwise undesirable consequences arise. On this level, we reveal a political application case of the study, which was made plausible with the help of an LCA. While the review of LCA studies was blindly divided up between the authors of this paper, in cases of uncertainty joint agreements were reached to ensure a consistent approach.

15.5 Results

The results of the study are rather sobering. Only 17 out of the 58 (equals 29%) studies indicate an application purpose (Fig. 15.1). The majority of these studies were carried out within the framework of ISO 14040/44. Another interesting observation is that, on the other side, although studies are carried out in accordance with the ISO standard they do not state the application purpose (in a sufficient manner). The general low naming of the application purpose did not allow further analysis of relations between the different attributes. Additionally, we investigated whether the publication media has an influence on whether the application is mentioned or not. Two observations could be made: papers in journals with a higher impact factor tend to name the application more often and papers in journals with a thematically broader scope and readership, respectively, also more often state the application purpose. This could indicate that information about the intended use and backgrounds of the work appeals to a wider audience and makes it more successful, as we will discuss later.

Nevertheless, there are huge differences in the assumptions, so we agree with the statement of van der Voet et al. (2010) that opaque assumptions have a huge impact on the results. However, we do not take these findings as the basis for the

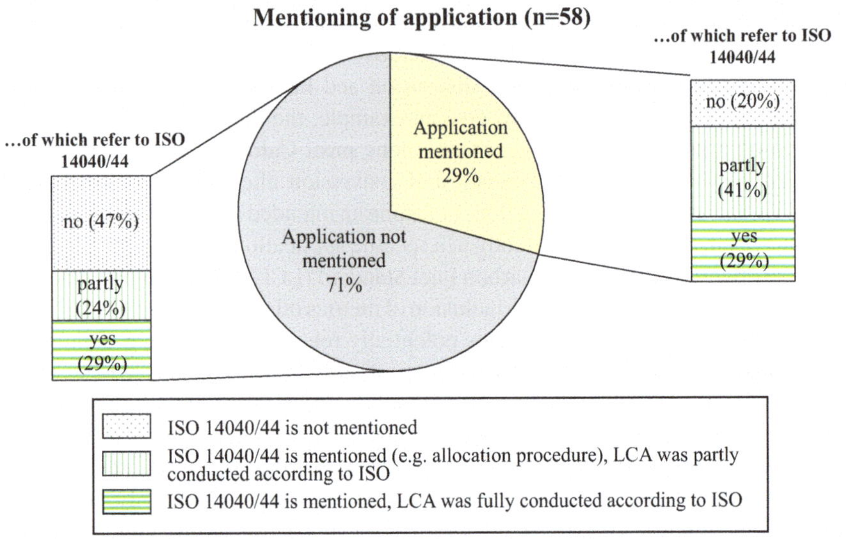

Fig. 15.1 A review of the application case is mentioned in the 58 LCA studies that served the base of the meta-study of (van der Voet et al. 2010)

demonization of the method, but instead, examine reasons for this observation and discuss perspectives for the present developments of life cycle assessment.

15.6 Discussion

The discussion of the results focuses on two levels: First, we argue that the application purposes in the studies were not apparent for various reasons so that the differences in LCA assumptions cannot be empirically justified by the application. On this basis, we propose to drop the concept of objectivity and self-purpose studies in order to focus more on the societal embedding of life cycle assessment and application purposes in research and teaching.

Despite training in life cycle assessment, engineering, and social sciences, we were not able to methodically determine the intended applications within the given corpus. Since our review was limited to a metastudy on biofuels only, further studies are needed to validate our findings and extend its scope to other fields of LCA. A wider corpus might reveal linguistic structures that facilitate clustering and categorization into application purposes and sharpen the review methodology. Nevertheless, we consider biofuels to be a vivid example for the discussion of the application purpose. While issues of food, water, energy, and mobility, and in particular, the focus on "tank versus plate", attract a wide audience, the life cycle assessments were a common reference point for different stakeholders. The considered studies and the societal debate impressively represent the methodological challenges and demand.

We, therefore, consider the critical reflection of the intended purpose of biofuel life cycle assessments to be particularly important and a good starting point for further studies.

With regard to methodological criticism, it is, on the one hand, possible that we were not able to find a purpose because we have mainly studied scientific publications, but industrial studies are rarely published. Our finding that articles in high ranked and wider addressing journals mention the purpose more frequently indicates that the target audience makes a difference. On the other hand, the authors of published LCA studies may even wish to keep the use of the study open or have no interest in practical application at all. This attitude may be further encouraged by the scientific ethos of disinterestedness and universality for the scientific publication, which forbids the indication of purposes (see Merton 1968, pp. 607–616). This idea could also have been reinforced by the community and journal specifications if the purpose was not explicitly asked for. Not mentioning the application purpose, however, only disguises but can not replace the justification for the assumptions. You do not even have to assume a manipulative intention. Studies are likely to be influenced by the habit of a chair or industry, by the available budget as well as the software learned and its technical limitations (see Wynne 1975; Lakatos and Musgrave 1974). Many decisions are made unconsciously or, respectively and, in the best intention of an individual for appropriate analysis of a system. But a lack of reflection on normativity and social consequences does not absolve us of our social responsibility. If there is no interest at all to say something about a target system, but only to play with and explore features of a toy-model, one could at least spell out this application purpose. In order to further reveal the application purpose, what we found in the written studies was not sufficient. Nevertheless, it would be possible to interview the authors of studies or to make a survey in a follow-up study. Alternatively, we would like to address the community directly to communicate the purpose of studies more strongly for the following reasons.

We could not show in our text analysis that there are connections between application purposes and assumptions of an LCA. We, therefore, assume that the reference to ISO 14040 is not sufficient to justify the results. However, we point out that the reflection and indication of the purpose of an LCA are helpful in justifying and making comprehensible assumptions in a study. The methodological development and the results of the meta-study show that LCA does not seem to converge to a consensus. Our theoretical consideration explains this diversification through the different societal environments of an LCA. The application of an LCA for a practical case is always a particular challenge. Mastering these practical problems, such as dealing with the data availability, different stakeholders and unclear objectives of the stakeholders, is another actual achievement of the studies. We, therefore, consider it important to communicate the handling of the application purpose in the studies so that other practitioners can learn from dealing with these problems. To do this, however, one must give up the idea of objective science, or unambiguous allocation of environmental impacts.

Objectivity or universality only make sense as the ideal to make the insights of a study independent from subjective impressions but transparent and comprehensible for a wider audience. We, therefore, claim that only the purpose of application relieves the modeler of subjective impressions, political bias and random circumstances in the justification of assumptions, considered input data, and simplifications. Therefore, not only the quantitative results of a study are the scientific achievement, but also the application of the general model to a concrete problem in a way that justifies assumptions, results and uncertainty assessments against the background of its social embedding, respectively the societal purpose of the application. Therefore, written instructions to the reader's imagination, for example in forms of narrative structures, are used to communicate relevant contextual knowledge and determine the framework in which the study was carried out. To illustrate particular problems, references to standards, methods, and stakeholders must be combined with observed interrelationships. The reasons for assumptions are then derived from this. Highlighting these multiple challenges would also open LCA for interdisciplinary exchange. With their help, studies could be geared more closely to the perspectives of stakeholders and find further commonalities in how assumptions in the creation of an LCA can be traced back to a list of application purposes.

15.7 Conclusion and Outlook

LCA is a success story, but with its success, there seems to be a crisis caused by the purpose-driven diversification. The scientific idea that the LCA studies might approach a consensus in methodology and results on e.g. a technology must be abandoned. But this is not a loss. There is no formal method to distinguish good science in other disciplines either but only heuristics and scientific virtues that guide the versatile practices (Lakatos and Musgrave 1974). Good engineers and scientists do not blindly follow rules given to them but solve cognitive and practical problems keeping in view, and in dialogue with, those affected by them. We, therefore, agree with the meta-studies that opaque assumptions strongly influence results. We conclude that mere reference to ISO standards or "the LCA method" does not justify the assumptions made and results obtained. And we go with the standards that LCAs are not an "end in themselves" but have a purpose. But we reveal that the application purpose can hardly be reconstructed (only in 29% of the studies). We, therefore, learn little about its applications and the associated peculiarities in the making and justification of assumptions. Thus, we can assume the worst case that the assumptions are arbitrary and manipulative. Then, LCAs are practically unusable in many cases (e.g. for policy and decision support). Or we consider the best case that the assumptions are based on particular definitions, practices, and purposes because LCAs are solutions for specific problems. Therefore, instead of the failed embedding into an (e.g. deliberative or strategic) practice, the LCA method was wrongly criticized. The intended purpose lies outside the standards that commonly define the LCA method, but is actually considered a part of every life cycle assessment study. The problem of LCAs is the lack of communication of the application purpose—on the one hand, as a prerequisite for the comparability of LCAs and, on the other hand, for the establishment of a culture that shares, discusses and appreciates revealed ways of solving practical problems. Our review uncovers and discusses this deficiency with the example of biofuels in order to encourage the reflection on the application purpose, especially in this field but also in the wider fields of life cycle assessment.

In view of the current global challenge of climate change, the comprehensive application of an LCA in industry, politics, jurisprudence, and science is considered to be important. There are experiences for successful projects so that not every study would require elaborate risk analysis or citizen participation. Nevertheless, these strategies play an important role in dealing with uncertainty in LCA—and are of varying importance for different application purposes (Heijungs and Huijbregts 2004). As mentioned above, tight legal standards are needed to create meaningful legal incentive systems or labels. At this point, a reference database with usual emissions would also make sense in order to easily check the plausibility of information provided by companies, for example, as there are standard rates for tax returns. From the point of view of consumer protection, it would also be helpful not to leave the eco-labeling to the economy. This works well in some areas, but not in all. Standardizations are only successful if they include different perspectives on feasibility, for example, concerning the cognitive and practical possibilities of consumers and audit institutions, without forgetting their purpose of environmental protection. In the political dimension, the policy level must be considered: for local projects, it is important to take the current situation and concerns of the public stakeholders into account. Discourse-analytical and participatory methods seem to be as necessary as robust assumptions and simple models. At a higher political level, for example, to advise the European Parliament, the Numeral, Unit, Spread, Assessment and Pedigree (NUSAP) system and the "post-normal-science" mindset that goes beyond the mere application of methods ("normal-science") have proved their benefit (Funtowicz and Ravetz 1990; van der Sluijs et al. 2005). The central idea of our criticism is to place the intended application at the center for consideration, selection, and justification of all assumptions and methods in the life cycle assessment. There are many approaches to consider different application purposes—and with this invitation, we provoke life cycle assessment to become the ultimate alphabet soup. But we expect a new structure of LCA exchange to emerge when the commonalities of studies with similar intended applications then become visible.

References

European Commission (2003) Communication from the commission to the council and the European parliament - integrated product policy - building on environmental life-cycle thinking: 52003DC0302. Retrieved from https://eur-lex.europa.eu/legal-content/EN/TXT/?uri=CELEX:52003DC0302

European Commission - JRC (2010) International reference life cycle data system (ILCD) handbook. General guide for life cycle assessment : detailed guidance. EUR 24708 EN. http://publications.jrc.ec.europa.eu/repository/handle/JRC48157. Accessed 18 July 2020

Klöpffer W, Grahl B (2012) Ökobilanz (LCA). Ein Leitfaden für Ausbildung und Beruf, 1., Auflage; Wiley-VCH: Weinheim. ISBN 3527659927

Hauschild MZ, Rosenbaum RK, Olsen SI (eds) (2018) Life Cycle Assessment. Theory and Practice. Springer International Publishing, Cham. ISBN 978-3-319-56475-3

DIN EN ISO 14040 (2009) 14040 Umweltmanagement – Ökobilanz – Grundsätze und Rahmenbedingungen, 2009, 13.020.10, 13.020.60 (14040)

von Falkenstein E, Wellenreuther F, Detzel A (2010) LCA studies comparing beverage cartons and alternative packaging: Can overall conclusions be drawn? Int J Life Cycle Assess 15:938–945. https://doi.org/10.1007/s11367-010-0218-x

Liu G, Müller DB (2012) Addressing sustainability in the aluminum industry: a critical review of life cycle assessments. J Clean Prod 35:108–117. https://doi.org/10.1016/j.jclepro.2012.05.030

Martin EW, Chester MV, Vergara SE (2015) Attributional and consequential life-cycle assessment in biofuels: a review of recent literature in the context of system boundaries. Curr Sustain/Renew Energy Rep 2:82–89. https://doi.org/10.1007/s40518-015-0034-9

Merton RK (1968) Social theory and social structure. Free Press, New York, NY

van der Voet E, Lifset RJ, Luo L (2010) Life-cycle assessment of biofuels, convergence and divergence. Biofuels 1:435–449. https://doi.org/10.4155/bfs.10.19

Yellishetty M, Ranjith PG, Tharumarajah A, Bhosale S (2009) Life cycle assessment in the minerals and metals sector: a critical review of selected issues and challenges. Int J Life Cycle Assess 14:257–267. https://doi.org/10.1007/s11367-009-0060-1

Guinée JB, Cucurachi S, Henriksson PJG, Heijungs R (2018) Digesting the alphabet soup of LCA. Int J Life Cycle Assess 23:1507–1511. https://doi.org/10.1007/s11367-018-1478-0

Finkbeiner M (2014) The international standards as the constitution of life cycle assessment: the ISO 14040 Series and its offspring: Chapter 3. In: Klöpffer W (ed) Background and future prospects in life cycle assessment. Springer, Dordrecht, pp 85–106. ISBN 9789401786973

Saltelli A, Giampietro M (2017) What is wrong with evidence based policy, and how can it be improved? Futures 91:62–71. https://doi.org/10.1016/j.futures.2016.11.012

Wardenaar T, van Ruijven T, Beltran AM, Vad K, Guinée J, Heijungs R (2012) Differences between LCA for analysis and LCA for policy: a case study on the consequences of allocation choices in bio-energy policies. Int J Life Cycle Assess 17:1059–1067. https://doi.org/10.1007/s11367-012-0431-x

Bauer C, Schebek L, Schmidt M (2007) Lebenszyklusanalysen und Entscheidungswissen. TATuP 16:10–16. https://doi.org/10.14512/tatup.16.3.10

Saltelli A (2019) A short comment on statistical versus mathematical modelling. Nat commun 10. https://doi.org/10.1038/s41467-019-11865-8

Walton KL (1990) Mimesis as make-believe. On the foundations of the representational arts; Harvard University Press, Cambridge, MA. ISBN 9780674576193

Toon, A (2014) Models as make-believe. Imagination, fiction and scientific representation. Palgrave Macmillan. ISBN 978-1-349-33687-6

Frigg R, Nguyen J (2016) Scientific representation. In: Zalta EN (ed) Stanford Encyclopedia of Philosophy

Luhmann N (1987) Soziale Systeme, 15th edn. Suhrkamp, Frankfurt am Main

Luhmann N (1986) Ökologische Kommunikation. Opladen, Westdeutscher

Luhmann N (1992) Die Wissenschaft der Gesellschaft, 1st edn. Suhrkamp, Frankfurt am Main

Nierling L, Torgersen H (2019) Normativität in der Technikfolgenabschätzung. TATuP 28:11–14. https://doi.org/10.14512/tatup.28.1.11

Grunwald A (2019) Technology assessment in practice and theory. Routledge, Abingdon, Oxon, New York, NY. ISBN 978-1-138-33708-4

Wynne B (1975) The rhetoric of consensus politics: a critical review of technology assessment. Res Policy 4:108–158

Finkbeiner M (2014) The international standards as the constitution of life cycle assessment: the ISO 14040 Series and its offspring: Chapter 3. In Klöpffer W (ed.) Background and future prospects in life cycle assessment. Springer, Dordrecht, pp 85–106. ISBN 9789401786973

DIN EN ISO 14044 (2018) Umweltmanagement – Ökobilanz – Anforderungen und Anleitungen, 13.020.10, 13.020.60 (14044)
Grunwald A (2016) Nachhaltigkeit verstehen. Arbeiten an der Bedeutung nachhaltiger Entwicklung. oekom verlag: München. ISBN 3865818218
Lakatos I, Musgrave A (eds) (1974) Kritik und Erkenntnisfortschritt. Vieweg, Braunschweig. ISBN 978-3-528-08333-5
European Commission (2013) Commission Recommendation of 9 April 2013 on the use of common methods to measure and communicate the life cycle environmental performance of products and organisations text with EEA relevance. 2013/179/EU. https://op.europa.eu/s/nDq4. Accessed 18 Dec 2019
Galatola M, Pant R (2014) Reply to the editorial "product environmental footprint—breakthrough or breakdown for policy implementation of life cycle assessment?": written by Prof. Finkbeiner (Int J Life Cycle Assess 19(2):266–271). Int J Life Cycle Assess 19:1356–1360. https://doi.org/10.1007/s11367-014-0740-3
Zhang Y, Joshi S, MacLean HL (2010) Can ethanol alone meet California's low carbon fuel standard? An evaluation of feedstock and conversion alternatives. Environ Res Lett 5. https://doi.org/10.1088/1748-9326/5/1/014002
Heijungs R, Huijbregts MAJ (2004) A review of approaches to treat uncertainty in LCA. http://www.iemss.org/iemss2004/pdf/lca/heijarev.pdf. Accessed 8 Feb 2017
Funtowicz SO, Ravetz JR (1990) *Uncertainty and quality in science for policy;* Kluwer: Dordrecht. ISBN 0-7923-0799-2
van der Sluijs JP, Craye M, Funtowicz S, Kloprogge P, Ravetz J, Risbey J (2005) Combining quantitative and qualitative measures of uncertainty in model-based environmental assessment: the NUSAP system. Risk Anal 25:481–492. https://doi.org/10.1111/j.1539-6924.2005.00604.x

Chapter 16
Biodiversity Impact Assessment of Grazing Sheep

Andreas Geß

Abstract Biodiversity is a complex and intangible field and a general applicable quantification methodology is yet to be established. Neither indicator systems on the influence on species richness or ecosystem services nor the assessment of the amount of human influence lead to a reliable quantification method. A new approach developed by Lindner in 2016 assesses the difference between a reference situation with the current situation through land use by combining expert knowledge with fuzzy logic and constructing trajectories of biodiversity impacts of case-specific indicator categories through interviews with a group of specialists. In this paper, the method is adjusted for the evaluation of the biodiversity impact of grazing sheep. A set of indicators was constructed, weighted and individually assessed together with sheep farming experts. The established indicators include the change of biodiversity through grazing sheep, the optimal grazed area, the optimal grazing period, the influences on the soil and the humus layer, the impact of machinery use, the importance of transhumance as well as the influence of other species like goats within the herd. An extensive sheep farm from the Swabian Alb, Germany and a semi-extensive sheep farm from the Rhön Mountains, Germany, were assessed for exemplarily testing the method. The assessment presented results assigning the more extensive sheep farm a higher biological as it was excepted. The set of indicators appears applicable for similar farms in similar regions. A point of further inquiry is to underline positive effects and benefits of grazing sheep on biodiversity observed in other studies.

Keywords Biodiversity · LCA · Lamb meat production

A. Geß (✉)
Life Cycle Engineering GaBi, University of Stuttgart, Institute for Acoustics and Building Physics IABP, Wankelstrasse 5, 70563 Stuttgart, Germany
e-mail: andreas.gess@iabp.uni-stuttgart.de

S. Albrecht et al. (eds.), *Progress in Life Cycle Assessment 2019*, Sustainable Production, Life Cycle Engineering and Management,
https://doi.org/10.1007/978-3-030-50519-6_16

16.1 Introduction

Biodiversity loss has become one of the major pressures on our planet. As the IUNC states, the amount of threatened species has doubled between 2000 and 2018 (IUNC 2018). Loss or further reduction of species would trigger a chain reaction and a decline of life forms higher up the food chain. Inevitably, the loss of biodiversity will affect us humans and our way of living. The preservation therefore is not only idealistic pastimes, it's vital for the holistic well-being of ourselves (Alcamo 2003).

Environmental scientists started addressing our current ecological and conservational situation as the "Anthropocene", which indicates that humans nowadays are the biggest changing force on ecology and geology (Corlett 2015). To counteract this development the anthropogenic production must be changed into a form that not only conserves our current biodiversity but actually contributes to the variability of species like for example grazing sheep dosuch as sheep promote. seed transport, overturn the upper soil layers and preserve clearings and pasturelands. According to various study outride these activities of sheep the industrial or machine based landscaping from an ecological and sustainable point of view (Härdtle et al. 2009; Pollock et al. 2013; Zerbe and Wiegleb 2009). This paper aims at developing an approach to quantify how biodiversity is influenced by grazing sheep farming.

Biodiversity has become an umbrella term for a variety of environmental issues. Definitions of biodiversity reach from species richness to ecological wellbeing. In 1992 the CBD defined biodiversity as the "the variability among living organisms from all sources including, inter-alia, terrestrial, marine and other aquatic ecosystems and the ecological complexes of which they are part; this includes diversity within species, between species and of ecosystems" (UNEP 1992). Regarding the multidimensional concepts of the term biodiversity, quantifying it has become a complex chore. For example species richness varies in every ecosystem and therefore it proved not applicable to use a single species as a representant for biodiversity as a whole. Each organism has different requirements and shows different reactions to changes (Birrer et al. 2014; Gabel et al. 2016; Stoeckli et al. 2017; Wilting and van Oorschot 2017). Other approaches, in which the biodiversity impact is evaluated through variables, are fitted for the examined industry or land use form, e.g. farming (Stoeckli et al. 2017). The issue that comes with an analysis of such a complex cause is the enormous effort for a single analysis of a certain production process or service, since the variables have to be adjusted for the region to the form of production or also local characteristics. Indicators and variables should represent all aspects of biodiversity. Consequently, a common agreement for assessing biological diversity has not been reached yet.

In 2008, Michelsen introduced a method that compares the current and the optimal state of an ecosystem and derives the biodiversity impact though the difference of the two (Michelsen 2008). Based on this approach, Linder developed a method to assess biodiversity through expert interviews and the use of fuzzy logic. It has already been adapted to the land use forms of forestry and cropland (Lindner et al. 2016; Perennes 2017). The aim of this study is to adjust Lindner's method to fit the local ecoregion and the land use form of livestock farming and hereby quantify to which extent grazing sheep influence the ecological community in two study areas on the Swabian Alb and the Rhön Mountains in Southern Germany.

16.2 Methodology and Study Areas

16.2.1 Methods

The study was conducted according to the UNEP/SETAC guidelines for the land use impact calculations in LCA which requires the creation of a spatial model before the data collection and the calculation of the land use impact (Koellner et al. 2013). Since the UNEP/SETAC guidelines have to be adjusted for sheep farming (see 4.2.1), the steps are modified as shown in Fig. 16.1

In 2016, Lindner developed a new method to build a framework for the integration of biodiversity in LCAs and find a way to put figures on the biodiversity impact of the various products. The method is based on using input from expert opinions and experience. Together with specialists in sheep farming, a questionnaire was put together to give a holistic and target-orientated overview about the biodiversity effects of the production of a specific product.

To calculate the biodiversity quality according to the UNEP/SETAC framework a rarity factor of the area in question is needed to weigh the value of the rarity of the appearing species of an Ecoregion. Thereto, the WWF Wildfinder database is used to estimate the rarity richness factor R_m of Ecoregion m. The biodiversity quality is calculated by multiplaying R_m by the biodiversity potential function μ_m, see formula

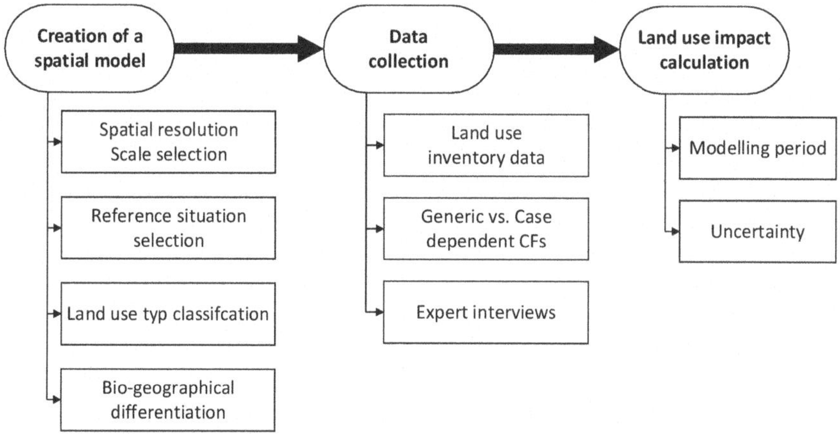

Fig. 16.1 Calculation procedure for the proposed land use impact assessment (adapted according to (Koellner et al. 2013; Saad et al. 2011))

(1). Its unit is the universal biodiversity unit (UBU), which is undefined and used for comparative reasons. The biodiversity potential function arises from the aggregation of the membership functions after linking and weighting of the single indicators. The weighting factors are developed together with the experts (Lindner et al. 2016).

$$Q_m = R_m *_m (\phi_i)\left[UBU/m^2 * a\right] \quad (1)$$

Q_m Biodiversity quality in Ecoregion m [UBU/m^2*a]
R_m Rarity Rated Richness of Ecoregion m
μ_m Biodiversity potential function [UBU/m^2*a]
ϕ_i Membership function of indicator i.

Since both study areas are located in the Ecoregion "Western European Broadleaf Forests", the same rarity richness factor of $R_{m,\,WEBF}$ 1.27 according to the WWF WildFinder database is accounted for both areas.

For the reference state of Q_m in these region $\mu_m = 1$ UBU/m^2*a, meaning that $Q_{m,}$ is equal to R_m hence $Q_{m,\,WEBF} = 1.27$ UBU/m^2*a.

The impact of the production process is assessed by following the UNEP/SETAC framework, and calculated by multiplying the difference of reference biodiversity quality and the process' biodiversity quality with the use of area per functional unit and the occupation time for the production of one functional unit with formula (2).

$$\text{Impact}_m = \Delta Q_m * A * t = R_m * (1 - \mu_m) * A * t\ [UBU] \quad (2)$$

A Area occupied for the production of a FU
t Occupation time for the production of a FU.

For the quantification of the multi-level concept biodiversity, a model that can deal with vague propositions is needed. Therefore fuzzy set theory is applied here. An element in fuzzy logic can have multiple aspects or can belong only gradually to a certain set (Zadeh 1965). In opposition to the Boolean theory or crisp set theory a truth value could not only be zero, meaning false, or one, meaning right (Boole 1847), but also everything in between. It is therefore a many-value concept and is able to grasp linguistic variables and quantify what cannot directly put in numbers (Saad et al. 2011; Yannis 2001; Zadeh 1990).

Fuzzy set theory makes use of similar operations as in crisp set theory. All sets and subset affiliations are defined by membership functions, μ(A)t. In crisp set theory, membership functions appoint the degree of belonging to either zero or one; they are so called step functions with clear boundaries, see Fig. 16.2. Membership functions of a fuzzy set can take any form, which implies that a crisp set membership function is a special case of a fuzzy set membership function (Center and Verma 1998).

In Lindner's method, biodiversity membership functions ϕ_i of the characterisation factors (CF), are adjusted and trajectory curves are created according to the expert's assumptions in relation to a reference state set by the expert him-/herself. The curves are later graphically summarized, averaged and the membership graph is adjusted to

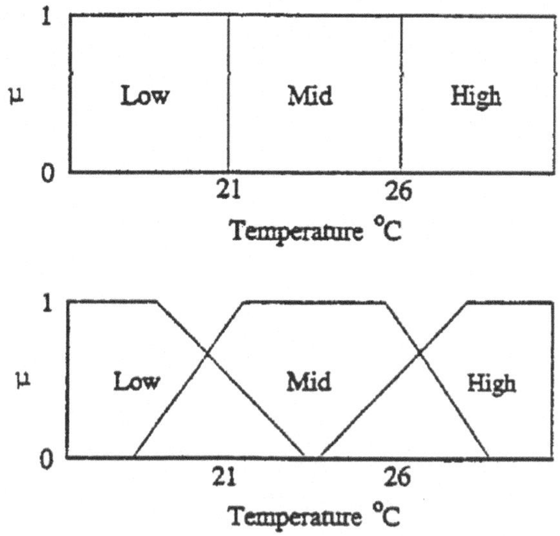

Fig. 16.2 Example of a graph for the membership functions in a crisp set with clear boundaries and a fuzzy set with overlapping boundaries (Center and Verma 1998)

the curve developed by the specialists. The basic membership function is a Gaussian function equation with six constants for adjustment, see formula (1). The graph of this function is displayed in Fig. 16.3, with an interval of [0; 1] for ϕ_i. For esthetic purposes the following constants have been chosen: $\alpha = 2$, $\beta = 0.5$, $\delta = 1$, $\varepsilon = 1$, $\gamma = 0$ and $\sigma = 0.15$.

$$y_i = \gamma + \varepsilon e^{\frac{|(x_i^\delta - \beta)^\alpha|}{2*\sigma^\alpha}} \, [-] \tag{1}$$

x_i	Management parameter
y_i	Biodiversity impact of the indicator x_i
α, β, δ, γ, ε and σ	Constants for adjustment.

The constants α, β, δ, γ, ε and σ modify the form of the graph so the graph can be adjusted to the expert's opinion. Through the hereby developed formula, it is possible

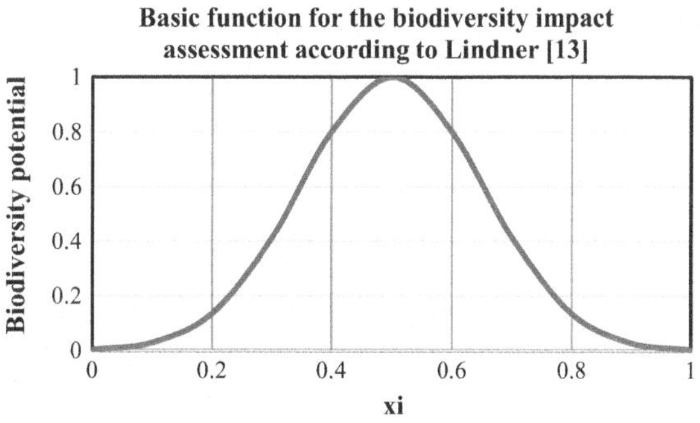

Fig. 16.3 Basic function of the biodiversity impact assessment with the constants $\alpha = 2$, $\beta = 0.5$, $\delta = 1$, $\varepsilon = 1$, $\gamma = 0$ and $\sigma = 0.15$

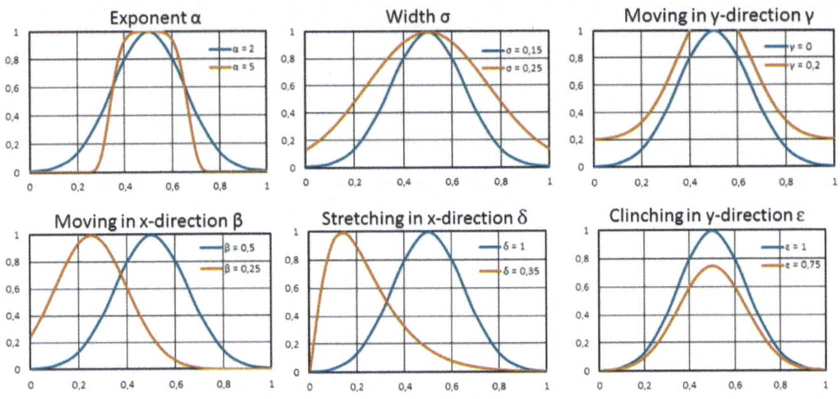

Fig. 16.4 Effect of the constants α, β, δ, γ, ε and σ according to Lindner, 2016 (Lindner et al. 2016)

to quantify the biodiversity impact of the examined indicator. The constant $\alpha \in \mathbb{Z}$ modifies the curvature of the graph around the vertex. The constant $\sigma \in \mathbb{R}$ alters the range of the curve and influences the curvature around the vertex. The constant $\beta \in \mathbb{R}$ moves the graph in x-direction, the constant $\gamma \in \mathbb{R}$ in y-direction. The constant $\delta \in \mathbb{R}$ moves the vertex in x-direction and compresses the graph on one side of the vertex and stretches it on the other. The constant $\varepsilon \in \mathbb{R}$ dsclinches or stretches the curve in y-direction (Lindner et al. 2016). A graphic explanation is given in Fig. 16.4.

A combination of constants is to be developed to fit the graph to the expert's opinions. The graph's formula serves as the basic function for the calculation of the regarded CF (Lindner et al. 2016).

The literature research was based on the literature quoted in Lindner, 2016 (Lindner et al. 2016), updated and extended by biodiversity studies of sheep farming. The interviews were conducted according to Lindner, 2016 (Lindner et al. 2016). After the literature research, a set of CF's was designed along with LCA-engineers experienced in ecology and biodiversity, landscape managers and consultants for ecological farming the different CF's and a ten-page questionnaire were iteratively evaluated, edited and examined on consistency and interdependence. The evaluation of the filled-out questionnaires was conducted anonymously. Eleven experts, independent from the study farms but familiar with the regional way of sheep farming, took part in the study, among them shepherds, sheep farmers, landscape managers, biologists and consultants for sheep and goat farming.

16.2.2 Study Areas

Although the two study areas belong to different nature regions—Münsingen to the Kuppenalb, Wildflecken to the Southern Rhön Mountains (Meynen et al. 1962)—and the endemic species differ slightly, they are regarded as the same type of region due to the similar altitude, climatic medium and the boundary condition with both being military training grounds. Furthermore, they both are located in the same Ecoregion called "Western European Broadleaf Forest" (Bohn et al. 2004). Hence, the developed questionnaire is applicable for both study areas.

The study farms' grazing fields are used as the spatial model in this case study. The farm in Münsingen comprises an area of 170 ha and the one in Wildflecken an area of 460 ha.

16.3 Results and Discussion

16.3.1 Characterization Factors

Through literature research and interviews with biologists, landscape managers and shepherds the essential points of the impact of grazing sheep on their surroundings have been worked out. The examined indicators are listed in the following:

- Change of biodiversity through grazing sheep over time
- Optimal grazed area per five hundred ewes
- Optimal grazing period throughout the year
- Change of soil by grazing sheep
- Change of the humus layer by grazing sheep
- Influence of the use of machinery on biodiversity
- Impact of grazing sheep on the nutrition cycle
- Impact of transhumance
- Impact of additional species inside a sheep flock.

All indicator trajectories are based on an exemplary sheep farm of 500 ewes under non-intensive management practices in south Germany. Also, no fertilizing of the pasture is assumed.

The developed trajectories for the assessment of the characterization factors are depicted in Fig. 16.5. For most CF's the appropriate curve was found. The experts suggested to also take the different vegetation and management types into account and to keep the process of evaluation simple to implement it into one of the CF's. For this reason the CF "Optimal grazing period throughout the year" shows for different curves, depending on the type of vegetation and the management of the pasture. The CF "Impact of additional species inside a sheep flock" was broken up into goats and donkeys. Aggregated a biodiversity impact potential of 1 is achievable, with goats reaching a maximum of 0.6 and donkeys a maximum of 0.4. A linkage of various indicators was not considered necessary.

According to the experts, the uncertainty of the individual graphs is the main challenge of the method. It was addressed through empiricism and using averaged curves. To assure the consent of all experts the averaged curves were shown them for revision.

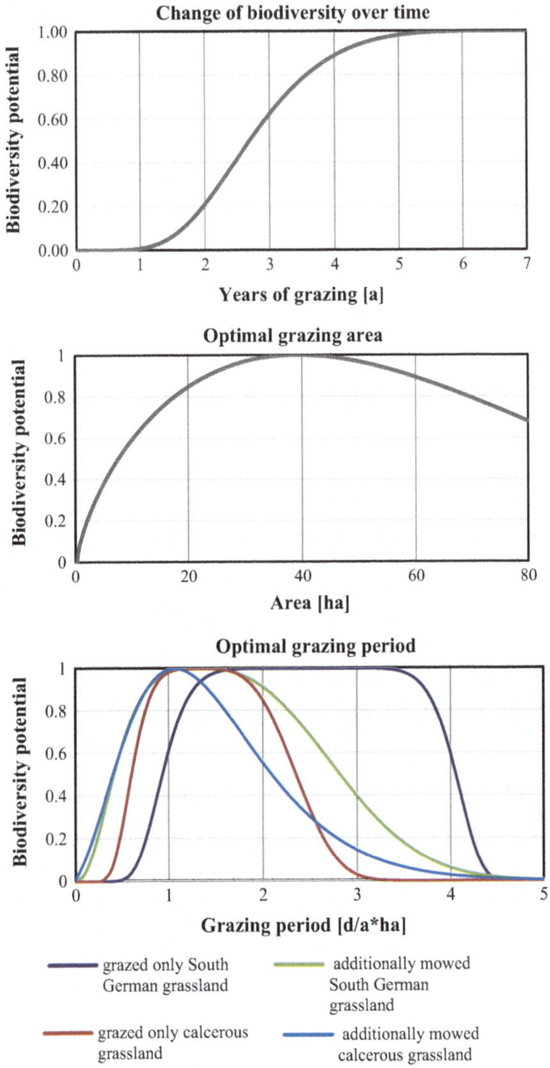

Fig. 16.5 Trajectories for the assessment if the biodiversity impact of the different indicators

16.3.2 Aggregation and Weighing

The weighing factors can be seen in Fig. 16.6. The highest values with a weighing of 0.2 out of 1.0 is reached in Change over time and Machinery use, while the impact on the humus layer was consider the lowest of the CF's.

The CF's were aggregated as the sum of all weighed CF's, resulting in a value between 1, meaning low impact on biodiversity or a high biodiversity quality, and 0, meaning a high impact on biodiversity or a low biodiversity quality.

Fig. 16.5 (continued)

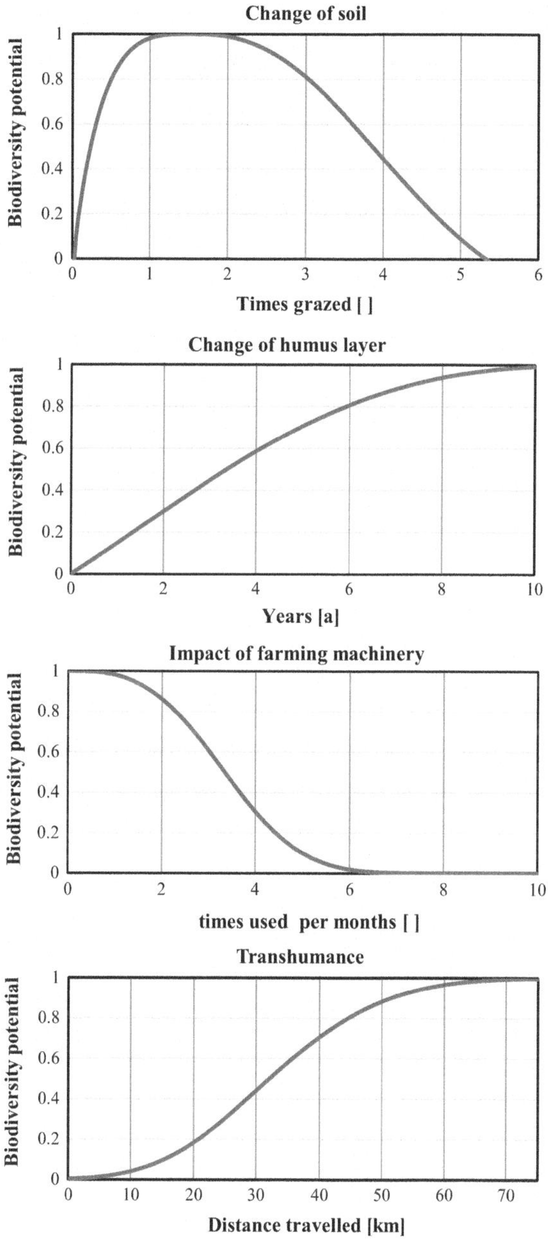

Fig. 16.5 (continued)

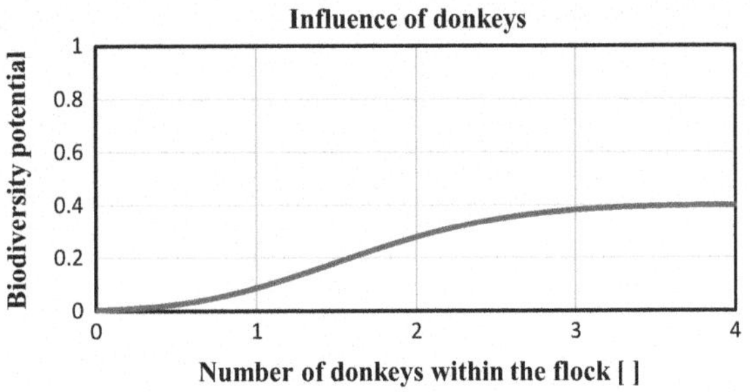

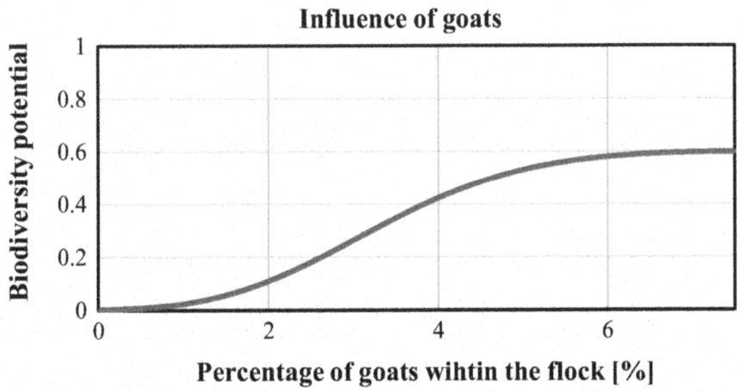

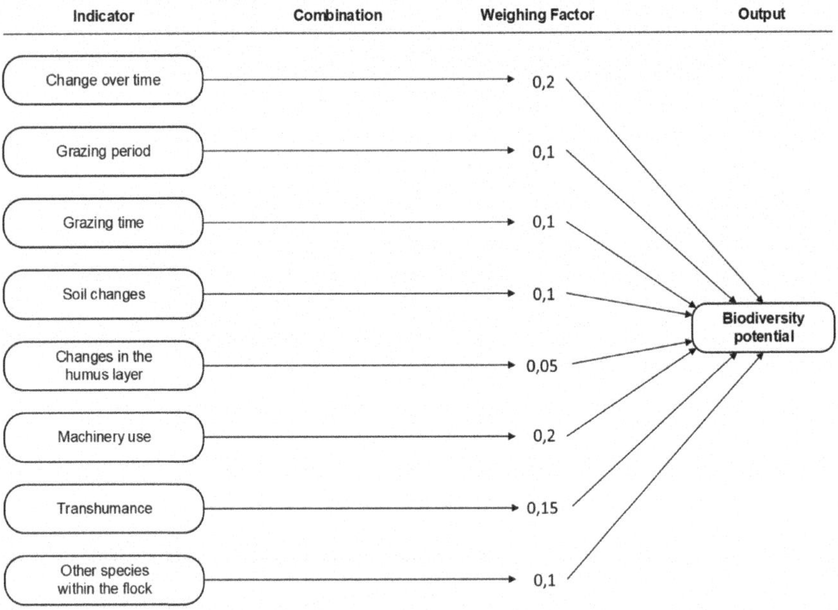

Fig. 16.6 Aggregation and weighing, in accordance with Perennes (Perennes 2017)

16.3.3 Results for the Study Farms

Table 16.1 shows the values for the different indicators of the two study farms. It shows that the key variations are in the grazing period and the distance of transhumance. In Fig. 16.7 the influence of these differences on the results is illustrated. Here, the biodiversity quality is calculated according to Eq. (1). For the Münsingen area a biodiversity potential μ_m of 0.90 resulting in a biodiversity quality of Q_m = 1.13 UBU/m^2*a and for the Wildflecken area a μ_m of 0.75 resulting in Q_m = 0.97 UBU/m^2*a was reached. The significant factors in this case study for the difference

Table 16.1 Indicator data for the calculation of the biodiversity potential of the study areas

Indicator data of the study farms			
Indicator	Unit	Farm Münsingen	Farm Wildflecken
Area grazed for	[a]	13	13
Grazing period	[d/ha]	1.4	0.8
Grazed area per 500 ewes per month	[ha]	21	23
Soil belayed for	[d/ha]	1.4	0.8
Humus layer affected for	[a]	13	13
Machinery use per month	[/30d]	1	2
Distance of transhumance	[km]	~120	~55
Influence of other species in the flock	[–]	1 donkey	~2% goats

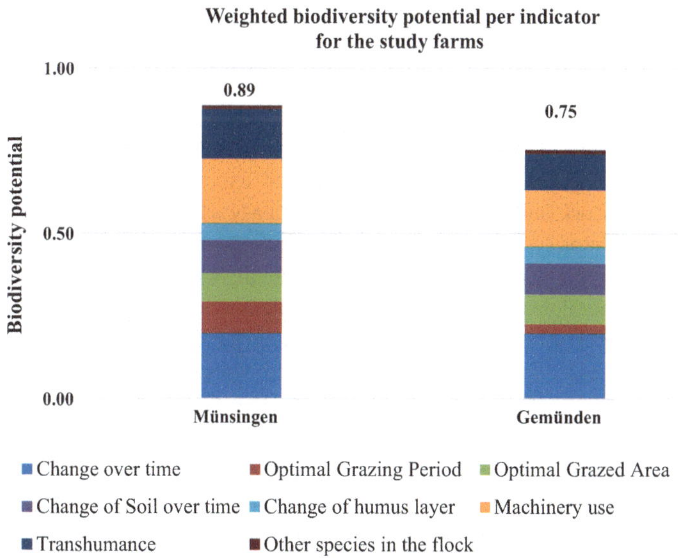

Fig. 16.7 Weighed Indicator values and total biodiversity potential of the study areas

in biodiversity potential are the transhumance and the grazing period. In Wildflecken transhumance is conducted only one way is therefore shorter and has a smaller potential than the one in Münsingen. The grazing period in Wildflecken has to be adjusted to the schedule of the military training ground. The larger area is necessary to be able to get around the daily training maneuvers.

In Lindner 2016, the impact per product is calculated according to the UNEP/SETAC guidelines (Lindner et al. 2016). Therefore, the biodiversity impact of 1 kg of lamb meat is calculated with Eq. (2). Since sheep eat around 4% of their bodyweight of dry matter of grass per day and the breeds at the two farms have an average weight of ca. 85 kg/ewe, the fodder demand is around 1.2 t of grass per year. The precise biomass production and therefore the feeding area demand per sheep can only be estimated and is controversially discussed with value for an average yield ranging from of 4.6 kg/m^2 (Statistisches Landesamt Baden-Württemberg 2017) to 5.6 kg/m^2 (Bayrisches Landesamt für Statistik 2016) resulting in demands between 214 and 261 m^2/sheep. In agricultural literature however, an area demand of 630 m^2 per ewe is defined (Beinlich and Krahl 1995). Also, the content of meat per lamb strongly depends on the climatic and conditions and therefore differs every season. Hence, the results for Münsingen are between 1.1 and 2.8 UBU/kg and for Wildflecken between 1.8 and 5.3 UBU/kg. Due to this missing reliability, the biodiversity impact of one kg of meat is not precisely assessable and the overall biodiversity potential of a sheep farms appears to be the more consistent result of study.

16.4 Conclusion and Outlook

The methodology proposed by Lindner in 2016 was adjusted for sheep farming. The assessment was developed within the UNEP/SETAC guidelines except for the assumption of a reference state, which was done individually by the experts. Dealing with the uncertainty of the individual curves appeared as the main challenge of the method. The advantage however is the inclusion of experienced knowledge from different fields of work and the relatively fast and easy assessment of further case studies once a indicator system is established. Both showed relatively little biodiversity impact whereas the more extensive farm is achieving a slightly better result. The study shows the general possibility of applying the method for sheep farming. Results can serve as a support in decision-making towards a for sheperds and consumers. The use of the indicator system is suitable for similar types of farms. For a generally valid application, a greater variety of study areas and experts are required. To achieve reliable results for the impact of one kg of meat it is necessary to conduct further inquiry concerning the biomass production per m^2 of pasture and the individual livestock fodder consumption. For the general evaluation of livestock farming, the example of grazing sheep shows a low representability, since in the study region all sheep farming is conducted in a similar, sustainable manner. For this matter further research is recommended.

References

Alcamo J (2003) Ecosystems and human well-being, Island, Washington, DC

Bayrisches Landesamt für Statistik (2016) Ernte der Feldfrüchte und des Grünlandes in Bayern 2015

Beinlich B, Krahl W (1995) Ein Naturschutzkonzept für die Kalkmagerrasen der Mittleren Schwäbischen Alb (Baden-Württemberg), Karlsruhe

Birrer S, Zellweger-Fischer J et al. (2014) Agric, Ecosyst Environ 197:195–203

Bohn U, Gollub G, Hettwer C, Neuhäuslová Z, Raus T, Schlüter H, Weber H, Hennekens S (2004) Map of the natural vegetation of Europe. Scale 1:2,500,000, Bundesamt für Naturschutz, Bonn

Boole G (1847) The mathematical analysis of logic. Macmillan Publishers, Cambridge, UK

Center B, Verma BP (1998) Artif Intell Rev 12(1/3):213–225

Corlett RT (2015) Trends Ecol Evol 30(1):36–41

Gabel VM, Meier MS et al (2016) J Environ Manage 181:249–260

Härdtle W, Assmann T et al. (2009) Renaturierung und Management von Heiden, Springer

IUNC (2018) IUNC red list statistics. http://www.iucnredlist.org/about/summary-statistics#TrendsInBiodiversityStatus, last Assessed: 28.08.2018

Koellner T, de Baan L et al (2013) The Int J Life Cycle Assess 18(6):1188–1202

Lindner JP, Sedlbauer K et al. (2016) Quantitative Darstellung der Wirkungen landnutzender Prozesse auf die Biodiversität in Ökobilanzen, Dissertation, Fraunhofer IBP; Fraunhofer IRB-Verlag

Meynen E, Schmithüsen J, Gellert JF, Neef E, Müller-Miny H, Schultze JH (1962) Handbuch der naturräumlichen Gliederung Deutschlands, Selbstverlag der Bundesanstalt für Landeskunde, Remagen und Bad Godesberg

Michelsen O (2008) Int J Life Cycle Assess

Perennes M (2017) Using local ecosystem indicators to determine land use impacts on biodiversity, A case study in Baden-Württemberg. Master thesis, Universität Bayreuth

Pollock ML, Holland JP et al (2013) Rangel Ecol Manag 66(4):387–400

Saad R, Margni M et al (2011) Assessment of land use impacts on soil ecological functions: development of spatially differentiated characterization factors within a Canadian context. Int J Life Cycle Assess 16(3):198–211

Statistisches Landesamt Baden-Württemberg (2017) Ernte der Hauptfeldfrüchte in Baden-Württemberg

Stoeckli S, Birrer S et al. (2017) Agric, Ecosyst Environ 237:224–233

UNEP (1992) United Nations Environmental Programme

Wilting HC, van Oorschot MMP (2017) J Clean Prod 156:194–202

Yannis A (2001) Phillis; Luc A. Andriantiatsaholiniaina. Ecol Econ 37(3):435–456

Zadeh LA (1965) Inf Control 8(3):338–353

Zadeh LA (1990) Internat. J Gen Syst 17(2–3):129–138

Zerbe S, Wiegleb G (2009) Renaturierung von Ökosystemen in Mitteleuropa. Spektrum Akademischer Verlag, Heidelberg

Correction to: Life Cycle Assessment of a Hydrogen and Fuel Cell RoPax Ferry Prototype

Juan Camilo Gomez Trillos, Dennis Wilken, Urte Brand, and Thomas Vogt

Correction to:
Chapter 2 in: S. Albrecht et al. (eds.), *Progress in Life Cycle Assessment 2019*, Sustainable Production, Life Cycle Engineering and Management, https://doi.org/10.1007/978-3-030-50519-6_2

In the original version of the book, the legend of the Fig. 2.2 was corrected. The chapter and book have been updated with the changes.

The updated version of this chapter can be found at
https://doi.org/10.1007/978-3-030-50519-6_2

S. Albrecht et al. (eds.), *Progress in Life Cycle Assessment 2019*,
Sustainable Production, Life Cycle Engineering and Management,
https://doi.org/10.1007/978-3-030-50519-6_17

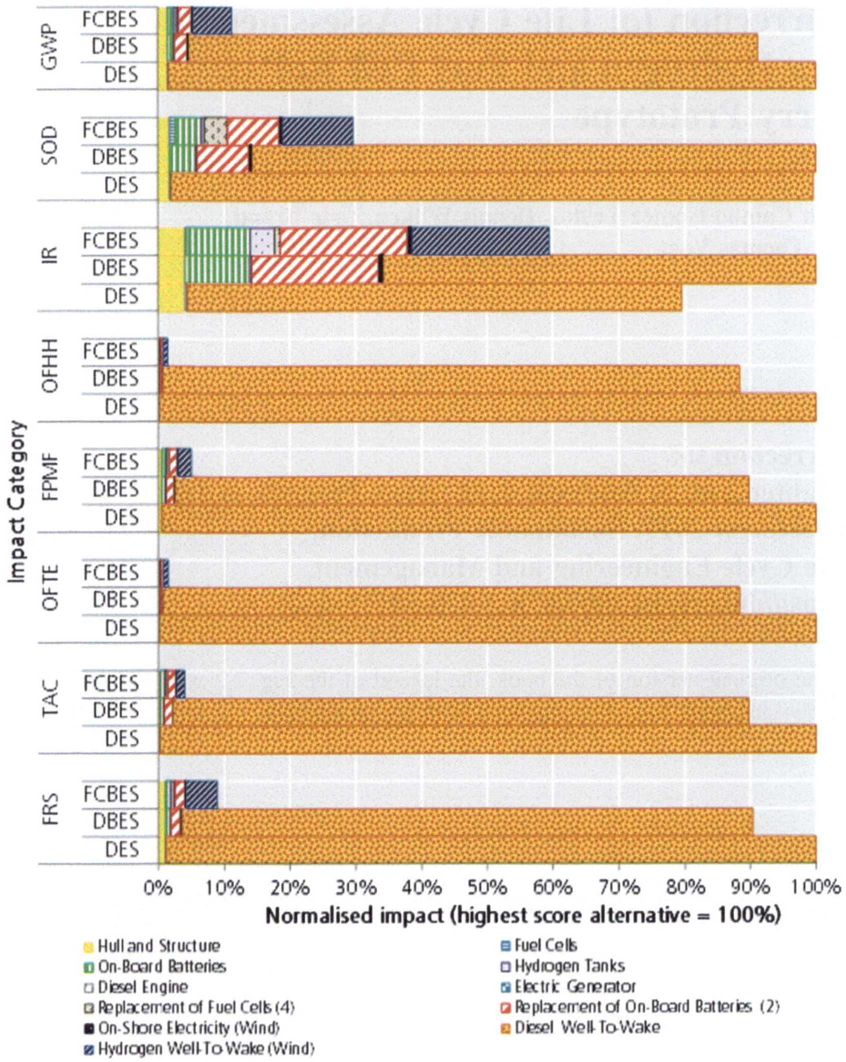

Fig. 2.2 Comparison of the midpoint impact assessment results for diesel electric ship (DES), diesel battery electric ship (DBES) and fuel cell battery electric ship (FCBES) RoPax ferry alternatives in the categories in which the FCBES has lower impact. Results normalised to the highest total impact alternative in each of the categories

The manufacturer's authorised representative in the EU is Springer Nature Customer Service Centre GmbH, Europaplatz 3, 69115 Heidelberg, Germany. If you have any concerns regarding our products, please contact ProductSafety@springernature.com

Printed and bound by CPI Group (UK) Ltd, Croydon, CR0 4YY
30/09/2026
02225503-0003